FLUID CRACKING CATALYSTS

CHEMICAL INDUSTRIES

A Series of Reference Books and Textbooks

Consulting Editor

HEINZ HEINEMANN

ADDITIONAL VOLUMES IN PREPARATION

FLUID CRACKING CATALYSTS

edited by

Mario L. Occelli
Georgia Institute of Technology
Atlanta, Georgia

Paul O'Connor
AKZO Nobel Catalysts
Amsterdam, The Netherlands

MARCEL DEKKER, INC.   NEW YORK · BASEL · HONG KONG

Library of Congress Cataloging-in-Publication Data

Fluid cracking catalysts / edited by Mario L. Occelli, Paul O'Connor.
   p. cm. — (Chemical industries ; 74)
   Includes bibliographical references and index.
   ISBN 0-8247-0079-1 (acid-free paper)
   1. Catalytic cracking. I. Occelli, Mario L. II. O'Connor, Paul. III. Series :
Chemical industries ; v.74.
TP690.4.F59   1997
665.5'33—dc21

                               97-43570
                                  CIP

The publisher offers discounts on this book when ordered in bulk quantities. For more information, write to Special Sales/Professional Marketing at the address below.

This book is printed on acid-free paper.

MARCEL DEKKER, INC.
270 Madison Avenue, New York, New York 10016
http://www.dekker.com

Current printing (last digit):
10 9 8 7 6 5 4 3 2 1

PRINTED IN THE UNITED STATES OF AMERICA

PREFACE

Since 1987, the petroleum division of the American Chemical Society has sponsored at three year intervals an international symposium on fluid cracking catalysts (FCC) technology. This volume reviews the progress in catalysts design and process innovation in an industry that, even 60 years after its introduction, still provides the main process of gasoline production worldwide.

Previous researchers have generated new families of cracking catalysts in response to federal and local environmental legislation and to the growing interest of refiners in cracking residual oils. Today refiners worldwide are meeting crude oil quality problems as well as environmental issues and regulations by developing, in cooperation with other scientists in industry and academia, new FCC formulations that will help improve air and water quality while meeting the challenges of producing reformulated gasoline.

Modern spectroscopic techniques are essential to the generation of detailed structural and composition analysis necessary to the understanding and advancement of the science of catalyst design and catalysis. Advances in catalyst design (and in the understanding of catalytic phenomena) closely parallel advances in modern spectroscopic techniques and other analytical instrumentation useful in identifying strategies to design new catalysts, correlate performance to surface properties, identify active sites, and, equally important, rationalize causes of catalyst deactivation. The use of solid state NMR, microcalorimetry, and atomic force microscopy to the study of FCC is described in several chapters. In addition, the reader will find that this book explores the effects of contact time on catalyst performance, introduces screening methods for predicting resid cracking activity, emphasizes the high activity and coke selectivity properties of FCC catalysts containing new metal tolerance matrices, reveals how inorganic acid treatments of ultrastable Y zeolite augment cracking performances, and describes special demetalization processing procedures that reactivate the spent FCC near its original equilibrium activity.

In conclusion, we would like to express our gratitude to colleagues everywhere for acting as technical referees and to Susan B. Occelli for her invaluable secretarial help and editorial assistance. The views and conclusions expressed herein are those of the chapter authors, whom we sincerely thank for their time and effort in presenting their research at the symposium and in preparing the camera-ready manuscripts for this book.

Mario L. Occelli
Paul O'Connor

CONTENTS

Preface

CONTRIBUTORS

Saeed Alerasool   Ashland Petroleum Company, Ashland, KY 41114

John M. Andresen   University of Strathclyde, Glasgow, United Kingdom

A. Auroux   CNRS, Villeurbanne, France

F. Baldiraghi   EURON, San Donato Milanese, Milan, Italy

J. R. Bernard   Elf Aquitane, Solaize, France

A. I. Biaglow   United States Military Academy, West Point, New York

Lori T. Boock   Grace Davision, Columbia, Maryland

H. Cauffriez   Institut Français du Pétrole, Rueil-Malmaison Cedex, France

Th. Chapus   Instiut Français du Pétrole, Rueil-Malmaison Cedex, France

Artie Chin   Mobil Technology Company, Paulsboro, New Jersey

S. Collet   Elf Aquitaine, CRES, Solaize, France

J. Coopmans   Akzo Nobel Catalysts, Amsterdam, The Netherlands

J. M. Dereppe   Université Catholique de Louvain, Louvain-La-Neuve, Belgium

Dilip J. Dharia   Stone & Webster Engineering Corporation, Houston, Texas

J. M. Domínguez   Instituto Mexicano del Petróleo, México D.F., Mexico

Patricia K. Doolin   Ashland Petroleum Company, Ashland, Kentucky

Maria D. Farnos   Mobil Technology Company, Paulsboro, New Jersey

M. Forissier   LGPC/CNRS-CPE Lyon, Villeurbanne, France

J. Fraissard   Université P. et M. Curie, Paris, France

R. J. Gorte   University of Pennsylvania, Philadelphia, Pennsylvania

S. A. C. Gould   Claremont College, Claremont, California

F. Hernández   Instituto Mexicano del Petróleo, México D.F., Mexico

James F. Hoffman   Ashland Petroleum Company, Ashland, Kentucky

L. H. Hsing   Phillips Petroleum Company, Bartlesville, Oklahoma

Ron Hughes   University of Salford, Salford, United Kingdom

J. C. Jansen   Delft University of Technology, Delft, The Netherlands

J. Kärger   Universität Leipzig, Leipzig, Germany

Xing-Min Ke   Research Institute of QiLu Petrochemical Co., Shandong, People's Republic of China

K. R. Kloetstra   Delft University of Technology, Delft, The Netherlands

Christopher W. Kuehler   Chevron Research & Technology Company, Richmond, California

S. Leoncini   EURON, San Donato Milanese, Milan, Italy

Warren S. Letzsch   Stone & Webster Engineering Corporation, Houston, Texas

Bing-Lan Li   Research Institute of QiLu Petrochemical Co., Shandong, People's Republic of China

Ch. Marcilly   Institut Français du Pétrole, Rueil-Malmaison Cedex, France

Brian J. McGhee   University of Strathclyde, Glasgow, United Kingdom

Abdul Aziz H. Mohammed   University of Strathclyde, Glasgow, United Kingdom

J. Navarrete   Instituto Mexicano del Petróleo, México D.F., Mexico

D. Nevicato   Elf Aquitaine, Solaize, France

P. Ngokoli-Kekele   Université P. et M. Curie, Paris, France

A. Nosov   Boreskov Institute, Novosibirsk, Russia

P. O'Connor   Akzo Nobel Catalysts, Amsterdam, The Netherlands

C. L. O'Young   Texaco Research and Development, Port Arthur, Texas

M. L. Occelli   Georgia Tech. Research Institute, Atlanta, Georgia

F. P. Olthof   Akzo Nobel Catalysts, Amsterdam, The Netherlands

Huifang Pan   Research Institute of QiLu Petrochemical Corporation, Shandong, People's Republic of China

Stephen K. Pavel   Coastal Catalyst Technology, Inc., Houston, Texas

A. W. Peters   Grace Davison, Columbia, Maryland

I. Pitault   LGPC/CNRS-CPE Lyon, Villeurbanne, France

Augusto R. Quinones *   Mobil Technology Company, Paulsboro, New Jersey

P. Rivault   Elf Aquitaine, Solaize, France

John Allen Rudesill   Grace Davison, Columbia, Maryland

R. Smeink   Akzo Nobel Catalysts, Amsterdam, The Netherlands

Colin E. Snape   University of Strathclyde, Glasgow, United Kingdom

M.-A. Springuel-Huet   Université P. et M. Curie, Paris, France

Jianming Su   Research Institute of QiLu Petrochemical Corporation, Shandong, People's Republic of China

E. Terrés   Instituto Mexicano del Petróleo, México D.F., Mexico

A. Toledo   Instituto Mexicano del Petróleo, México D.F., Mexico

H. van Bekkum   Delft University of Technology, Delft, The Netherlands

Jin-Shan Wang   Research Institute of Qilu Petrochemical Co., Shandong, People's Republic of China

G. D. Weatherbee   Grace Davison, Columbia, Maryland

Geoffrey L. Woolery   Mobil Technology Company, Paulsboro, New Jersey

Xing-Zhong Xu   Research Institute of Qilu Petrochemical Co., Shandong, People's Republic of China

G. Yaluris   Grace Davison, Columbia, Maryland

H. W. Zandbergen   Delft University of Technology, Delft, The Netherlands

Xinjin Zhao   Grace Davison, Columbia, Maryland

*Currrent address: Akzo-Nobel Chemicals, Research Center, Pasadena TX

FLUID CRACKING CATALYSTS

1996 FLUID CATALYTIC CRACKING UPDATE

Warren S. Letzsch
FCC/DCC Program Manager
Dilip J. Dharia
Manager, FCC Technology & Business Development

STONE & WEBSTER ENGINEERING CORPORATION
1430 Enclave Parkway
Houston, Texas 77077

INTRODUCTION

Fluid Catalytic Cracking is continuing to evolve to meet the challenges facing refiners. The reaction system is becoming very specialized in order to produce a more specialized and varied yield slate. High-efficiency stripping and gentle, effective regeneration are important to operations that are run under more severe conditions and/or use customized catalysts.

1

Improved process design and more advanced catalysts are making resid fluid cracking the preferred route for upgrading the 650° F-plus portion of the crude barrel.

Some of the more recent FCC developments are highlighted in this paper.

FEEDSTOCK INJECTION

The Stone & Webster/IFP feed injection system has become the standard to which other systems are compared. Since its inception in the early 1980's, it has offered refiners the most economical way to significantly upgrade their FCC operation. In Table 1, typical improvements observed in FCC operations with improved feed injection technology are shown. Lower dry gas and delta coke yields translate directly into higher throughput or conversion while the bottoms cracking always improves. Typical payouts are 2 - 4 months.

The feed injection technology has advanced considerably over the years even though the basic design looks very much the same. These changes are highlighted in Table 2. A continuous testing program has led to a better understanding of the design and operation of the entire system. While a high oil pressure (about 150 psig) ensures good atomization and the flexibility to

Typical Feedstock Injection Improvements
Table 1

Dry Gas	Down	10-40%
LPG	Down	0-10%
Gasoline	Up	1.5-8.0%
Bottoms Cracking		Maximized
Delta Coke	Down	0.05-0.15

Stone & Webster/IFP
Feed Injection Improvements
Table 2

1982	High Pressure Impact Atomizer
1983	Riser Coverage (Spray Pattern)
1985	Catalyst Preparation
1987	Moderate Pressure Design
1988	Low Pressure Impact Design
1989	Nozzle Size Options
1991	Improved Flux Control
1995	Higher Efficiency Design

run heavy residual feeds, many systems have been designed for low viscosity feedstocks. These nozzles require relatively low pressures (50 psig) and less atomization steam than residual applications.

A new design is being introduced this year. It is an impact nozzle that is significantly more efficient than the older design. This is illustrated in Figure 1, which shows that typical droplet sizes are reduced about 20% . Smaller droplets are always beneficial for achieving the kind of yield improvements previously shown.

Mechanical reliability, the hallmark of the Stone & Webster design, has not been sacrificed in the latest configuration. Competitive designs are much more complex, have small orifices and rely on internals to atomize the feed. These

SWEC Feed Injection Nozzle
Figure 1

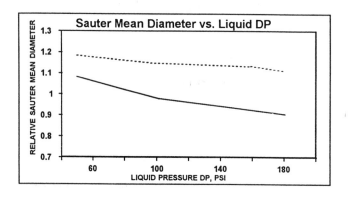

nozzles are very difficult to service and are subject to severe erosion and orifice plugging both from solids which are usually present in the feed and also from scale that may come in with the steam. Run lengths of 3 - 4 years are unlikely without sacrificing performance prior to the scheduled shutdown.

Other problems caused by poor nozzle design are plugging and coking. The former is more apt to happen with low pressure designs, while coking is associated with low steam rates. This coke can form in or around the nozzle, in the riser, or on the walls of the transfer line to the main fractionator. The more complex the nozzle design, the more likely a plugged or coked nozzle will result.

Attached Table 3 summarizes the performance to date with the Stone & Webster/IFP technology. Over 1.7 million barrels of cat feed have been licensed, an amount which far exceeds all the newer competitive designs. No other feed system can match the simplicity, performance, maintenance record, versatility and experience of the Stone & Webster/IFP feed system.

POST-RISER REACTION CONTROL

The secondary cracking that occurs in the dilute phase of the reactor vessel is undesirable since it destroys the valuable products. Both thermal cracking and

Feedstock Injection Technology
Table 3

- 48 Operational and Committed Designs

- Over 1,700,000 B/D Capacity

- Highest Reliability

- Lowest Maintenence

hydrogen transfer reactions occur in the dilute phase and produce dry gas from

LCO and gasoline and cause a loss in olefinicity (C_2 -C_4's). There are two

sections to the post-riser cracking: (1) the dilute phase and (2) the time in the

separators and overhead system leading to the bottom of the main fractionator.

Since thermal cracking is the main reaction occurring, both contact time and

temperature are the main reaction variables.

Stone & Webster's technology approach is to control both variables and yet

keep the operation mechanically simple, easy to operate, and minimize the

chance of any catastrophic occurrence that could shutdown the FCC operation.

While hard-connecting the cyclones eliminates the dilute phase, it creates a host of other problems. These include:

- Rigid start-up and operational procedures

- Main fractionator catalyst carryover

- Soft coke in the dilute phase

- Complex mechanical design

- Increased time for reactor heat-up during start-ups

In place of the closed-coupled concept, a system that addresses all of the above problems has been designed. A Ramshorn terminator (or axial cyclone) has been substituted for the primary cyclone, and a short surge space has been incorporated to keep pressure or circulation surges from filling the main fractionator with catalyst. The results from several commercial operations have generally shown a 15 - 20 % reduction in dry gas, low catalyst losses, and easy unit start-ups. Both full and half Ramshorn designs are operational, and other configurations are being modeled.

A vapor quench system is also offered as an alternative technology or in conjunction with the advanced axial cyclone system. Quench technology works by lowering the reaction temperature downstream of the feed riser. The temperature drop lowers the thermal cracking taking place in the hardware and dilute phase. When used with the Ramshorn technology, the quench gives the lowest dry gas yield. This combination is particularly attractive when reactor temperatures are high ($\geq 990°F$). The capital cost required for quench systems is normally quite modest, and the technology has been proven in several FCC operations.

RESID DESIGN

In order to run resid effectively, several technologies can be applied. These include two- stage regeneration, reaction mix temperature control and catalyst coolers.   Two-stage regeneration is a must. It protects the catalyst from hydrothermal deactivation by removing the moisture at low temperature in the first stage, while completing the regeneration in the dry atmosphere of the second stage. Staging the regeneration also permits heat removal from the first stage via the CO produced. For a 30,000 B/D unit, this can be the equivalent of a 70 MM BTU/HR catalyst cooler, as illustrated in Figure 2. Metals effects are minimized with this design because it keeps moisture away from the catalyst when it is at the highest regeneration temperature. It is well known that fluid cracking catalysts barely deactivate at all in a dry atmosphere at 1500°F. The

Two Stage vs. Single Stage Regeneration
Figure 2

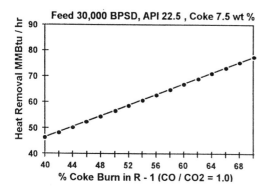

combination of vanadium, water and high temperature is particularly detrimental as the resulting

$$2VO + 3H_2O \rightarrow 2VO(OH)_3 \text{ (vanadic acid)}$$

vanadic acid attacks the catalyst structure and causes a permanent activity loss.

Other designs use a single-stage regenerator with a catalyst cooler to process resid. Here unwanted heat is generated and then has to be removed. The high moisture and metal contents environment is particularly rough on the catalyst, so the operational strategy usually proposed is to run the cooler to drop the regenerator temperature to 1300°F or less. This produces excess coke and virtually assures all of the Conradson Carbon will quantitatively go to coke since the catalyst temperature (the vaporization medium) is lower than the feed endpoint.

Keeping the total catalyst inventory low is desirable from an operating viewpoint since it makes catalyst activity and selectivity more of a true operating variable. Since most of the catalyst inventory is in the regenerator, this inventory should be minimized without designing a system that is harmful to the catalyst. The objective is not to see how fast the carbon can be burned but to ensure the catalytic properties are retained even in the presence of feed-added contaminants such as sodium and vanadium. High catalyst/air mix temperatures are particularly detrimental to catalyst stability when the delta coke is high.

REACTION MIX TEMPERATURE CONTROL

Reaction mix temperature control is shown in Figure 3. Cracked naphtha is recycled downstream of the feed injection zone. Because the naphtha makes very little delta coke, it has the effect of raising the mix temperature of the catalyst and feedstock at the feed injection point and helps vaporize the heavier components in the feed. Attached Figure 4 shows the new temperature profile

Mixed Temperature Control
(MTC) Technology
Figure 3

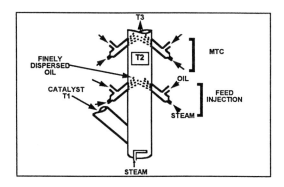

Riser Temperature Profiles (MTC)
Figure 4

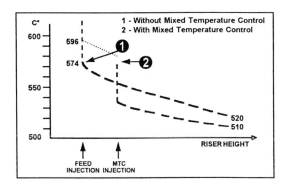

in the feed riser. Higher catalyst-to-oil ratios also result since mix temperature control increases the coke yield slightly while reducing the delta coke.

CATALYST COOLERS

Dense phase catalyst coolers are suitable when the Conradson Carbon goes above 7. These have been the most unreliable method of controlling the heat balance due to leaks which necessitate a unit shutdown or the removal of the heavier feed components. There are two main aspects to the design of the coolers: the heat transfer rate and the reliability. The former dictates the size of the cooler while the latter determines the run length or overall economics. The cooler design makes no difference to the FCC operation; it is simply the heat removal capacity that affects unit performance. Here ease of operation and turn-down capabilities are assumed to be similar. Since the process is really relying on the operability of a heat exchanger, Stone & Webster offers a design that will work even when a leak might occur. Tube modules that are individually piped allow a leak to be isolated on-stream with only a marginal (2 - 5 %) loss of heat removal capacity. Figure 5 shows the basic design of such a system. The experience with this design is quite extensive, as illustrated in Table 4.

Modular Flow Through Catalyst Cooler
Figure 5

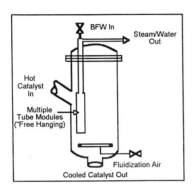

Catalyst Cooler Experience
Table 4

Number of Coolers in Operation	14
Number of Coolers under Design	2
Feedstocks	Atm. & Vac. Resids
Duty, MMBTU/Hr	8 to 140+
Steam Pressure, PSIG	175 to 850
First Installation	1983
Years of Operation	92+

PROCESS OUTLOOK

The FCC process remains the most versatile and important one in the refinery. It can handle the widest range of feedstocks and contaminants and produce the most varied yield slate. The Deep Catalytic Cracking process is an example of the adaptability of the technology to new demands. Further refinements in the catalysts and process will allow heavier feeds to be run and make residual processing the standard of FCC operations. Likely advances include shorter reaction contact times at higher temperatures in new riser configurations. Catalysts are expected to be more metals-tolerant and more sophisticated in the future. As an equipment designer and supplier, Stone & Webster will continue to lead the way in offering technical packages that can take maximum advantage of these catalyst improvements.

[AL,B]-ZSM-11 FCC ADDITIVE PERFORMANCE

L. H. Hsing and C. L. O'Young

Texaco Research and Development, 4545 Savannah at Highway 73,
Port Arthur, TX 77641

A Texaco R&D developed [Al,B]-ZSM-11 additive was evaluated on the Miniature Riser unit to study its effectiveness in promoting light olefin production, particularly branched olefins, for downstream alkylation, oxygenate processes, and results were compared with those for [Al]-ZSM-5.

[Al,B]-ZSM-11 additive increased C3=, C4=, and C5= yields, and was less effective in increasing C3=, C4= yields, particularly I-C4= but was more effective in promoting C5 olefin production than [Al]-ZSM-5.

The data imply that the acidity of [Al,B]-ZSM-11 additive is significantly lower than that of [Al]-ZSM-5.

Thermal deactivation of [Al,B]-ZSM-11 additive lowered its capability of promoting C4 (particularly IC4=) production but had little effect on C5 olefin yields.

The Clean Air Act Amendments of 1990 (CAAA) mandate a significant change in gasoline composition to meet increasingly stringent environmental emission requirements. it is anticipated that an FCCU in the refining industry will undergo a significant change from a major gasoline producer to an important olefin generation unit to meet future reformulated gasoline needs. Several options are available to the refineries to achieve the goals, which include the hydrotreating of FCC feedstock or naphtha products. Among the options available to the FCCU, the use of [Al]-ZSM-5 additive in an FCCU is one option available to the refinery, and has been practiced in the industry to increase olefin production. However, commercially-available [Al]-ZSM-5 additive suffers the deficits of lower gasoline yield as well as promoting more propylene instead of more valuable butylene and pentylene production for alkylation, MTBE, and TAME processes. Therefore, there is an incentive to develop an alternative additive with improved performance in promoting C4 and C5 olefins, particularly iso-olefins production.

EXPERIMENTAL WORK

The feedstock used came from a Texaco Refinery, and its properties are given in Table 1. An FCC equilibrium catalyst was used to carry out the work. The properties of the equilibrium catalyst are listed in Table 2.

The [Al,B]-ZSM-11 additive used was spray-dried by PQ Corporation, and the zeolite used in the additive was developed by Texaco R&D. The properties of the additive are listed in Table 3. [Al,B]-ZSM-11 is similar in pore structures to [Al]-ZSM-5, but with a different channeling direction and exhibits the same shape-selective characteristics and acidity as [Al]-ZSM-5.

Table 1. Feedstock Test Results

API gravity	21.4
Pour, F	91
Aniline point, F	163
Bromine no.	16.6
Watson aromatics, wt%	60.8
X-ray sulfur, wt%	2.517
Basic N2, WPPM	412
Total N2, WPPM	1949
Asphaltene, wt%	0.07
Vis K. 40C	71.57
Vis K. 100C	7.69
Vis K. 76.7C	14.96
R. I. 70C	1.4974
Pentane insolubles, wt%	3.8
N-C7 insolubles, wt%	0.08
Micro carbon residue, wt%	0.680
	D1160, F
IBP	546
5	645
10	680
20	723
30	761
40	805
50	834
60	868
70	905
80	950
90	1003
95	1046
EP	1078

Table 2. Catalyst Test Results

Catalyst	Equilibrium Catalyst
Activity, MAT (Davison)	69
Metals on catalyst	
Ni, wppm	270
V, wppm	700
Fe, wt%	0.54
La, wt%	1.2
Ce, wt%	0.24
Na2O, wt%	0.47
Compositions	
Al2O3, wt%	35.4
SiO2, wt%	59.1
Surface Area (BET), m^2/gm	
Total	153
Matrix	50
Zeolite	103
Pore Volume, cc/gm	0.36
Bulk Density, gm/cc	0.846
Unit Cell size, A	24.31

Incorporation of boron into the framework of ZSM-11 will weaken the acid strength of the zeolite. The acidity of [Al,B]-ZSM-11 has been characterized by TPD and FTIR techniques (1,2). The framework composition of the [Al,B]-ZSM-11 additive is 42.5% of Si, 0.14% of B, and 0.041% of Al. The additive was also thermally-deactivated in the presence of nitrogen at 1500°F for 8 hours to simulate the equilibrium additive.

The performance data were obtained on a miniature circulating riser pilot unit (MRU). A simplified flow diagram of the unit is shown in Figure 1.

Results and Discussion

Conversion and yields for all the runs with and without [Al,B]-ZSM-11 additive are plotted in Figures 2-17. The use level for fresh additive was 3 wt% whereas for thermally-deactivated additive was 4 wt%. Yield data with 4 wt% [Al]-ZSM-5 additive from a previous report (3) are also plotted and shown in the figures for comparison.

Table 3. [Al,B]-ZSM-11 Additive Properties

Thermal Deactivation	No.	1500F for 8 hours
Compositions, wt%*		
VALFOR CP711 Sieve	40	40
Clay	40	40
Alumina	20	20
Particle Size Distribution, %*		
<4.7 micron	0.0	—
4.7–6.6	0.0	—
6.6–9.4	0.1	—
9.4–13	0.2	—
13–19	1.0	—
19–27	3.7	—
27–38	13.9	—
38–53	9.1	—
53–75	19.2	—
75–106	15.9	—
106–150	10.8	—
150–212	19.8	—
212–300	6.6	—
Particle Size at 50%, micron*	84	—
% Attrition Loss, 0-19 micron*	46	—
Surface Area, m^2/g		
Zeolite	126	121
Matrix	109	74
Total	235	195

*Provided by vendor.

    As shown in Figure 2, propylene yield increased with the addition of either fresh or deactivated [Al,B]-ZSM-11 additive and the rate of yield increment for both [Al,B]-ZSM-11 is approximately the same. However, the propylene yield increase with the addition of [Al,B]-ZSM-11 additive is significantly lower than that for [Al]-ZSM-5 additive, indicating lower acidity of [Al,B]-ZSM-11. This is in agreement with the general consensus that the incorporation of boron into the framework of the Al-zeolite structure will weaken the acid strength of the zeolite.

    Isobutylene and normal butene yields are given in Figures 3 and 4, which shown IC4- and N-C4- yields increasing with the addition of [Al,B]-ZSM-11 additive. Again, the C4 olefin yields increment with [Al,B]-ZSM-11 additive is lower than that for [Al]-ZSM-5 additive. The yield ratios in Figures 5, 6, and

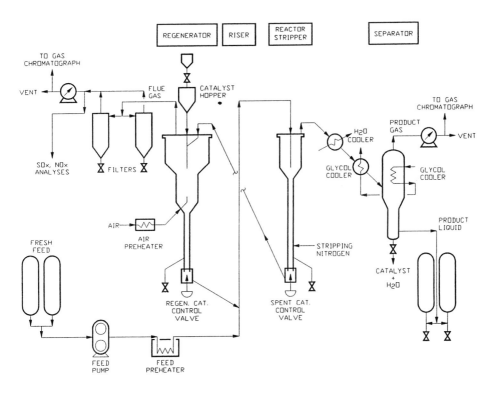

FIGURE 1. MINIATURE RISER FCC PILOT UNIT

7 indicate that the production of C4 olefins was lower for [Al,B]-ZSM-11 than [Al]-ZSM-5 at 960°F and 1000°F riser outlet temperatures. This is particularly true for isobutylene due to lower acidity of [Al,B]-ZSM-11 additive and is supported by the IC4=/total C4= yield ratio shown in Figure 8 that no apparent increment in the yield ratio was obtained with the addition of [Al,B]-ZSM-11 additives. This indicates that isomerization of normal butene to isobutylene requires acidity higher than [Al,B]-ZSM-11 additive. Yield ratio is defined as yields with and without additive at constant riser outlet temperature.

Branched and linear C5 olefin yields are shown in Figures 9 and 10, and their respective yield ratios are given in Figures 11 and 12. As shown in Figures 9 and 10, yields increased for branched C5 olefins while normal C5 olefin yields decreased with the addition of [Al,B]-ZSM-11 additives. This is inconsistent with the C5 olefin yields with the addition of [Al]-ZSM-5 in a previous report (3), which increased branched C5 and decreased normal C5 olefin yields. No

Figure 2
C3= Yield vs Conversion

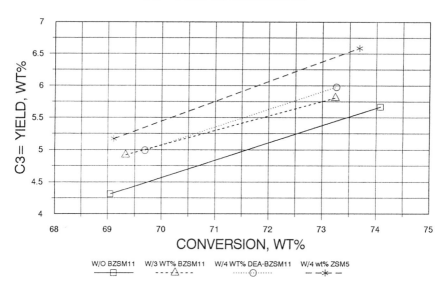

Figure 3
I-C4= Yield vs Conversion

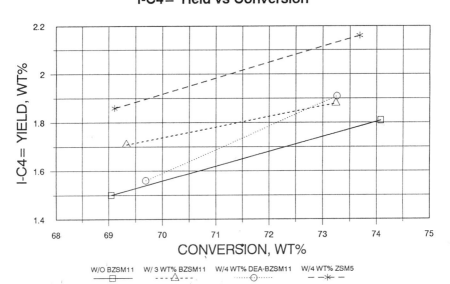

Figure 4
N-C4= Yield vs Conversion

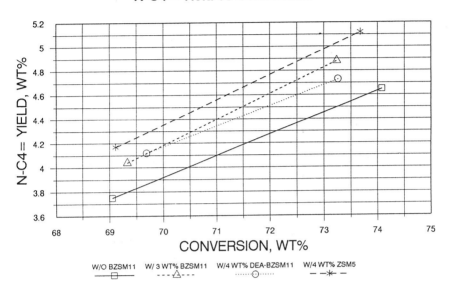

Figure 5
IC4= Yield Ratio vs Riser Outlet Temperature

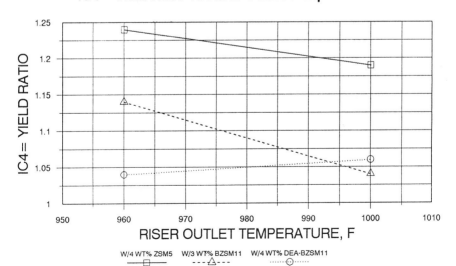

YIELD RATIO :
YIELD WITH/YIELD WITHOUT ADDITIVE

Figure 6

NC4= Yield vs Riser Outlet Temperature

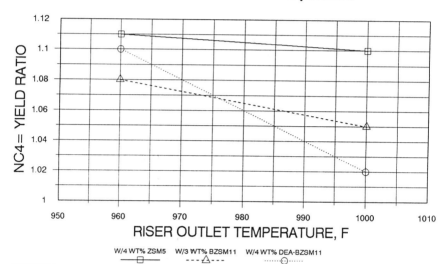

YIELD RATIO :
YIELD WITH/YIELD WITHOUT ADDITIVE

Figure 7
Total C4= Yield Ratio vs Riser Outlet
Temperature

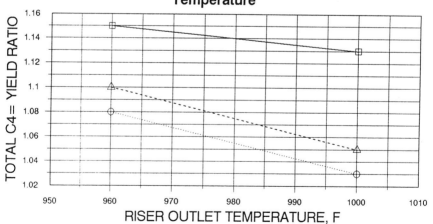

YIELD RATIO :
YIELD WITH/YIELD WITHOUT ADDITIVE

Figure 8
IC4=/Total C4= vs Riser Outlet Temperature

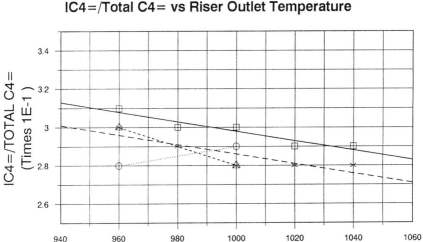

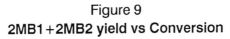

Figure 9
2MB1+2MB2 yield vs Conversion

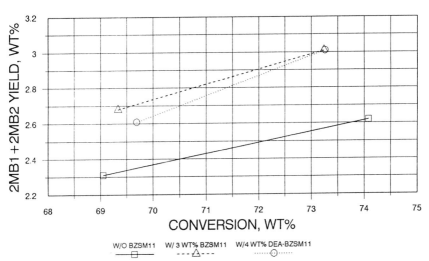

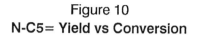

Figure 10
N-C5= Yield vs Conversion

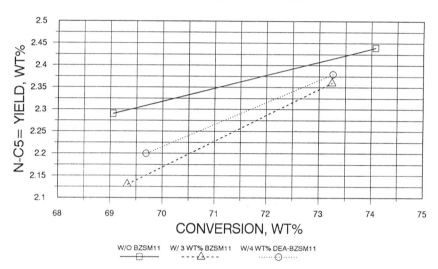

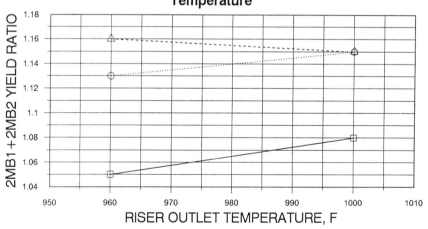

appreciable yield difference in branched and normal C5 olefins, shown in Figures 9 and 10, was noted for 3 wt% fresh and 4 wt% thermally-deactivated [Al,B]-ZSM-11 additives. C5 olefin yield ratios in Figures 11 and 12 indicate that both branched and normal C5 olefin yield ratios for [Al,B]-ZSM-11 additives are higher than those for [Al]-ZSM-5. This suggests that the acidity of [Al]-ZSM-5 is significantly stronger than that of [Al,B]-ZSM-11, and tends to crack more C5 olefins to lighter components.

Total C5 olefin yield ratio, shown in Figure 13, is higher for [Al,B]-ZSM-11 than [Al]-ZSM-5 additive, and again, confirms the stronger acidic characteristics of [Al]-ZSM-5 in comparison with [Al,B]-ZSM-11. It is also noted that C5 olefin yield ratio in Figure 13 is greater than 1.0 for [Al,B]-ZSM-11, whereas the yield ratio is less than 1.0 for [Al]-ZSM-5. This is because the cracking rate from heavier components to C5= is faster than C5-cracking to lighter components for [Al,B]-ZSM-11, while the opposite is true for [Al]-ZSM-5 additive. It is interesting to note that the C5= yield ratio is exactly the same for 3 wt% fresh and 4 wt% thermally-deactivated [Al,B]-ZSM-11 additives. Branched C5=/total C5= yield ratio in Figure 14 indicates that the yield ratio is the highest for [Al]-ZSM-5, followed by [Al,B]-ZSM-11, and is the lowest without the additive. The lower (2MB1+2MB2)/total C5= for [Al,B]-ZSM-11 suggests a lower skeletal isomerization of C5 due to lower acidity, although a higher total C5 yield and yield ratio was obtained for [Al,B]-ZSM-11 than for [Al]-ZSM-5. Nevertheless, [Al,B]-ZSM-11 additive promotes some skeletal isomerization of normal C5= to branched C5= indicating that skeletal isomerization of normal C5= requires lower acidity than that for C4 olefins.

The results reported above indicate that the additive with high acidity, resulting in high cracking, tends to promote lighter olefin (C3, C4) production at the expense of higher olefin components (C5+) in the gasoline range. High cracking is normally accompanied by high skeletal isomerization which, in turn, determines the iso/total olefin ratio. The acidity required for C5= skeletal isomerization is lower than that for C4=. The additive with stronger acidity such as [Al]-ZSM-5 is not necessarily good since it will lower gasoline and C5= yields and produce high C3 olefins. Therefore, an additive with a proper balance between cracking and isomerization is needed to optimize C3-C5 olefin production as demonstrated here. This is further supported by the data in Figure 15 that higher (2MB1+2MB2)/IC4= is found for [Al,B]-ZSM-11 than [Al]-ZSM-5, and [Al,B]-ZSM-11 is more suitable for TAME production while [Al]-ZSM-5 is favored for MTBE process.

Naphtha yield and yield ratio are given in Figures 16 and 17. As shown in Figure 17, naphtha yields decreased with the addition of [Al,B]-ZSM-11 additives. As shown in Figure 17, the naphtha yield ratio was higher for [Al,B]-ZSM-11 additives than for [Al]-ZSM-5 indicating less naphtha components

Figure 12
N-C5= Yield Ratio vs Riser Outlet Temperature

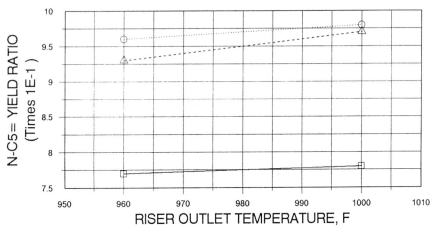

W/4 WT% ZSM5   W/3 WT% BZSM11   W/4 WT% DEA-BZSM11
   ——□——   - - -△- - -   ·······◯·······

YIELD RATIO :
YIELD WITH/YIELD WITHOUT ADDITIVE

Figure 13
Total C5= Yield Ratio vs Riser Outlet
Temperature

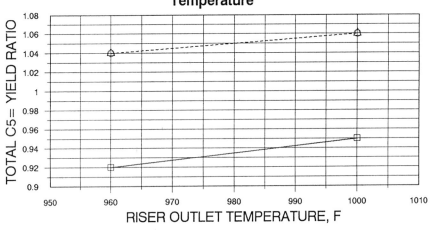

W/4 WT% ZSM5   W/3 WT% BZSM11   W/4 WT% DEA-BZSM11
   ——□——   - - -△- - -   ·······◯·······

YIELD RATIO :
YIELD WITH/YIELD WITHOUT ADDITIVE

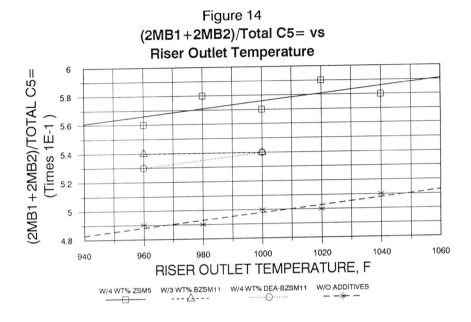

Figure 14
(2MB1+2MB2)/Total C5= vs
Riser Outlet Temperature

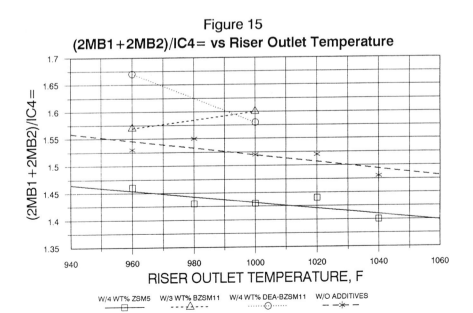

Figure 15
(2MB1+2MB2)/IC4= vs Riser Outlet Temperature

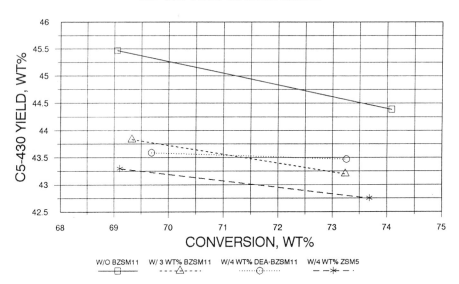

Figure 16
C5-430 Yield vs Conversion

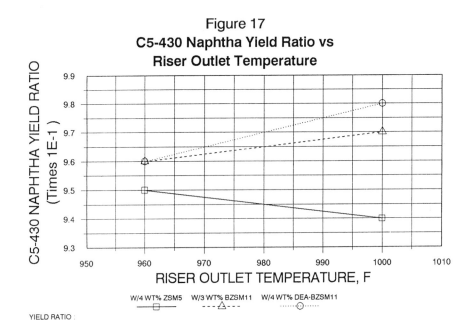

Figure 17
C5-430 Naphtha Yield Ratio vs
Riser Outlet Temperature

YIELD RATIO :
YIELD WITH/YIELD WITHOUT ADDITIVE

cracking to lighter products for [Al,B]-ZSM-11 additive. This is consistent with the lower acidity of [Al,B]-ZSM-11 additives.

CONCLUSIONS

1. [Al,B]-ZSM-11 increased C3, C4 production but the yield increase was lower than that for [Al]-ZSM-5.
2. [Al,B]-ZSM-11 increased C5 production and the yield increment was higher than that for [Al]-ZSM-5.
3. [Al,B]-ZSM-11 promoted very little skeletal isomerization for n-C4=.
4. [Al,B]-ZSM-11 promoted n-C5 skeletal isomerization but at a lower rate than that for [Al]-ZSM-5.
5. [Al,B]-ZSM-11 decreased naphtha yield less than [Al]-ZSM-5.
6. [Al,B]-ZSM-11 is more effective in increasing C5 and less effective in promoting C4= production than [Al]-ZSM-5 for oxygenate and alkylation processes.
7. The data imply that acidity of [Al,B]-ZSM-11 is much lower than that of [Al]-ZSM-5.
8. Thermal deactivation of [Al,B]-ZSM-11 additive lowered its C4 olefins, particularly IC4=, promotion capability but had very little effect on C5 olefin yields.

ACKNOWLEDGMENTS

We would like to express our appreciation to Texaco, Inc. for allowing the publication of the paper.

REFERENCES

1. Simon, W. M.; Nam, S. S.; Xu, W. Q.; Suib, S. L.; Edwards, J. C.; O'Young, C. L., *J. Phys. Chem., 96*, 6381–6388 (1992).
2. Bianchi, D. L.; Simon, M. W.; Nam, S. S.; Xu, W. Q.; Suib, S. L.; O'Young, C. L., *J. Cata., 93*, 551–560 (1993).
3. Hsing, L. H.; Pratt, R. E., Presented at the American Chemical National Meeting, August 24, 1993, and published in the ACS Symposium Series 571, November 1994.

THE EFFECT OF SILICA-TO-ALUMINA RATIO ON THE PERFORMANCE OF ZSM-5 FCC ADDITIVES

Christopher W. Kuehler
Chevron Research and Technology Company
100 Chevron Way
Richmond CA 94802

INTRODUCTION

FCC additives containing ZSM-5 are widely used in the petroleum refining industry to increase gasoline octane and produce light olefins. Although some progress has been made in improving the activity of additives through chemical stabilization and improved accessibility, little has been done to modify selectivities.

In many cases, an FCC is restricted from enjoying the octane benefits of ZSM-5 because it is limited in its ability to recover the products, in particular, the light gases and LPG. It has been discovered[1,2] that as the silica-to-alumina ratio of the ZSM-5 crystal increases, the incremental ratios of octane/LPG, octane/gasoline loss, and C4=/C3= increase. In this paper, pilot plant data is provided which substantiates these observations, and more detailed changes in gasoline composition are discussed.

How ZSM-5 Works

Despite a number of excellent studies on ZSM-5, there remains a widespread impression that ZSM-5 increases octane by cracking straight chain, low octane paraffins out of the gasoline boiling range. Buchanan[3] clearly demonstrated the low reactivity of these paraffins relative to olefins and both he and Madon[4] presented excellent explanations of the mechanism.

This mechanism involves the cracking of normal and branched olefins out of the gasoline boiling range to form propylene and butylenes before they can undergo hydrogen transfer to low octane paraffins on the host catalyst. ZSM-5 has also been demonstrated to isomerize olefins. Octane is increased through increased yields of C_5 hydrocarbons, an increase in aromatics concentration, a reduction in yields of C_{7+} paraffins and straight chain olefins, and an increase in the branched/normal ratio for olefins and paraffins.

The rates of the various reactions catalyzed by ZSM-5 vary with the carbon number of the reactant. The rate of cracking for C_5 olefins is low but increases rapidly with carbon number. Olefin isomerization, on the other hand, proceeds much more rapidly than cracking for C_{5-} reactants and either remains the same or decreases with increasing carbon number[8]. For heavier olefins in the gasoline, the rates may become comparable. Thus factors effecting the ability of ZSM-5 to catalyze each of these reactions, such as aging or initial silica-to-alumina ratio could affect their relative rates and influence the reaction products and gasoline composition.

EXPERIMENTAL

ZSM-5 Additives

All of the additives used in the study were prepared with approximately 11 wt% ZSM-5 crystal in a silica-alumina commercial binder. The ZSM-5 crystal came from several sources and was characterized only by their silica-to-alumina ratio. Based on previous unpublished experience with "equilibrating" additives prepared with this binder, no steaming was employed. This work indicated that steaming of additives in this binder had little effect on either activity or selectivities. If there is any "equilibration" that does occur, it must occur in the few hours the additive experiences in the pilot plant prior to the yield period.

Three additives were used in the study. The first, typical of FCC octane additives commonly used, contained a zeolite with a silica-to-alumina ratio of 40. The silica-to-alumina ratios of the ZSM-5 crystal in the other two additives were nominally 550 and 850. As the silica-to-alumina ratio of the crystal increases, the precision of the measurement suffers due to the very low aluminum content. In addition, although we could discern no significant effect of steaming on the pilot plant performance of the additives, there is still the expectation of some dealumination as evidenced by changes in selectivities as the additive ages commercially[5,6,7]. It should be noted that this observation is based on early additive formulations and that the binder chemistry of modern formulations may inhibit this dealumination. Nevertheless, it would be expected that these silica-to-alumina ratios represent the lower bound of the value. An accurate measurement of the zeolite silica-to-alumina ratio in the additive is difficult because of the presence of alumina in the binder.

FCC Riser Pilot Plant Runs

The additives were blended with equilibrium catalyst to give 8% ZSM-5 in the blend. Although high compared to typical commercial operation, this level was chosen to ensure large enough process effects to be able to accurately differentiate between the additives. The equilibrium catalyst contained a nominally zero rare earth, USY zeolite. Matrix activity was high. Inspections are given in Table 1.

TABLE 1

Base Catalyst Inspections

Physical Properties

ABD, g/cc	0.843
PBD, g/cc	0.9185
Attrition, DI	1.4
PSD, Wt %	
0-20 Microns	0.5
0-40 Microns	3.9
0-80 Microns	52.6
0-125 Microns	91.9
APS	78.2
UCS, Ang.	24.23

Chemical Composition, Wt %

Total REO	0.14
SiO_2	56.8
Al_2O_3	42.2

The feed used was hydrotreated vacuum gas oil with Arabian Heavy and Maya origins. Inspections are given in Table 2.

The experimental data was taken on a circulating riser pilot plant at a feed rate of 1400 g/hr. Catalyst inventory was approximately 6000 g. Riser outlet temperature was maintained at 950°F. Catalyst circulation was controlled with an auger device and measured by blocking flow with a flapper valve and measuring the pressure drop of the accumulating catalyst. The temperature profile in the 50 ft. riser was set by an adiabatic control scheme. Constant riser outlet temperature was maintained by varying external heat input at the base of the riser.

Data was acquired at two cat-to-oil ratios. Typically there was no effect of the ZSM-5 additive on conversion, however, minor variations in operating conditions did affect conversion at times. Having data at two conversions allowed adjustment to constant conversion. The pilot plant was typically lined

TABLE 2

Feed Inspections

Gravity, API	24.6
Aniline Point, °F	184.9
Sulfur, Wt %	0.35
Total N, ppm	957
Basic N, ppm	202
Rams Carbon, Wt %	N/A
Microcarbon Residue, Wt %	0.19
TBP, °F, Wt %	
St	585
5	656
10	684
30	750
50	806
70	869
90	968
95	1018
EP	1100
Recovered	

out for two to three hours at one set of conditions, a 1 1/2 hour yield period was taken, conditions were adjusted to the second circulation rate and lined out, then a second 1 1/2 yield period was taken. Yield periods of this duration allow sufficient product for distillation and subsequent analysis for properties such as engine octanes. For the purposes of simplicity in this paper, the results reported are differentials from the base catalyst for both yields and product properties.

RESULTS

Yield and product property shifts from the base catalyst(no additive) are given in Table 3. Light and heavy gasoline compositional shifts are given in Tables 4 and 5, respectively. Liquid yields are based on chromatographic analysis as

TABLE 3

Pilot Plant Yield and Octane Shifts

	8% 40 SiO$_2$/Al$_2$O$_3$ ZSM-5		8% 550 SiO$_2$/Al$_2$O$_3$ ZSM-5		8% 850 SiO$_2$/Al$_2$O$_3$ ZSM-5	
	Wt %	Vol %	Wt %	Vol %	Wt %	Vol %
Hydrogen	-0.0		-0.0		-0.0	
Methane	-0.1		+0.0		-0.0	
Ethane	-0.1		+0.0		-0.0	
Ethylene	+0.3		+0.0		+0.0	
C$_2$-	+0.1		+0.0		-0.0	
Propane	+0.4	+0.8	+0.2	+0.4	+0.0	+0.0
Propylene	+3.5	+6.1	+1.3	+2.3	+0.7	+1.2
Isobutane	+1.1	+1.8	+0.2	+0.4	-0.0	-0.1
N-Butane	+0.2	+0.3	+0.1	+0.1	-0.0	-0.0
Butylenes	+3.3	+4.9	+1.3	+2.0	+1.0	+1.5
C$_5$-265°F (LCN)	-4.4	-5.8	-0.2	-0.2	-0.1	-0.1
265-430°F (HCN)	-3.7	-4.3	-2.0	-2.5	-1.2	-1.5
C$_5$-430°F (Whole)	-8.1	-10.0	-2.2	-2.6	-1.3	-2.1
430-540°F (LCO)	-0.5	-0.5	-0.0	-0.0	-0.2	-0.2
540-650°F (MCO)	+0.1	+0.0	+0.2	+0.2	+0.1	+0.0
650°F+ (HCO)	+0.4	+0.4	-0.2	-0.3	+0.1	+0.1
Coke	-0.5		-1.1		-0.3	
LCN RON	+2.8		+1.6		+1.3	
LCN MON	+0.8		+0.6		+0.5	
HCN RON	+4.0		+1.5		+1.6	
HCN MON	+2.9		+0.9		+1.5	
Whole RON	+3.3		+1.6		+1.4	
Whole MON	+1.7		+0.7		+0.9	
Whole (R+M)/2	+2.5		+1.1		+1.2	

TABLE 4

Shifts in Light Gasoline Composition, Vol %

		Saturates				Olefins				Aromatics	Dienes	Uncl.	Total
		Iso	Normal	Naph	Total	Iso	Normal	Naph	Total				
40 SiO₂/Al₂O₃	C₅	+1.96	+0.38	+0.09	+2.43	+7.04	+1.00	+0.12	+8.16		+0.05	+0.00	+10.64
	C₆	-0.20	+0.13	+0.16	+0.09	+1.39	-0.88	+0.45	+0.96	+0.75	+0.02	+0.00	+1.82
	C₇	-2.28	+0.08	+0.39	-1.81	-1.86	-1.49	-0.42	-2.93	+1.10	+0.00	+0.00	-3.64
	C₈	-2.23	-0.50	-2.07	-4.80	+0.00	+0.00	+0.00	+0.00	-0.57	+0.00	-1.08	-6.45
	C₉	-0.83	+0.00	-0.22	-1.05	+0.00	+0.00	+0.00	+0.00	+0.00	+0.00	-0.23	-1.28
	Total	-3.58	+0.09	-1.65	-5.14	+6.57	-1.37	+0.99	+6.19	+1.28	+0.07	-1.31	
550 SiO₂/Al₂O₃	C₅	-0.28	+0.06	-0.32	-0.54	+2.59	+0.18	+0.04	+2.81	+0.00	+0.01	+0.00	+2.28
	C₆	+0.56	+0.01	+0.15	+0.72	+1.29	-0.30	+0.04	+1.03	+0.09	+0.00	+0.00	+1.84
	C₇	-0.62	-0.29	+0.27	-0.64	-0.78	-0.70	-0.04	-1.44	-0.02	+0.00	+0.00	-2.10
	C₈	-0.63	-0.18	-0.57	-1.38	+0.00	+0.00	+0.00	+0.00	-0.07	+0.00	-0.27	-1.72
	C₉	-0.05	+0.00	-0.01	-0.06	+0.00	+0.00	+0.00	+0.00	+0.00	+0.00	-0.05	-0.11
	Total	-1.02	-0.40	-0.48	-1.90	+3.10	-0.82	+0.12	+2.40	-0.00	+0.01	-0.32	
850 SiO₂/Al₂O₃	C₅	+1.05	-0.01	-0.20	+0.84	+3.47	-0.24	+0.01	+3.24	+0.00	+0.02	+0.00	+4.10
	C₆	-0.48	+0.02	-0.05	-0.51	+2.14	-0.69	+0.09	+1.54	+0.01	+0.01	+0.00	+1.05
	C₇	-0.64	-0.11	-0.12	-0.87	-0.40	-0.85	+0.09	-1.16	-0.22	+0.00	+0.00	-2.25
	C₈	-0.71	-0.20	-0.87	-1.78	+0.00	+0.00	+0.00	+0.00	+0.09	+0.00	-0.61	-2.30
	C₉	+0.09	+0.00	-0.01	+0.08	+0.00	+0.00	+0.00	+0.00	+0.02	+0.00	+0.04	+0.14
	Total	-0.69	-0.30	-1.25	-2.24	+5.21	-1.78	+0.19	+3.62	-0.10	+0.03	-0.57	

TABLE 5

Shifts in Heavy Gasoline (HCN) Composition, Vol %

	Saturates				Unsaturates				Aromatics	>200°C	Total
	Branched	Normal	Cyclic	Total	Branched	Normal	Cyclic	Total			
8% 40 SiO$_2$/Al$_2$O$_3$ ZSM-5											
C$_7$	0.10	0.02	0.14	0.26	0.00	0.00	0.02	0.02	0.43		0.72
C$_8$	0.71	0.15	2.04	2.90	0.12	-0.16	0.57	0.53	3.15		6.58
C$_9$	-1.65	0.00	0.17	-1.48	-1.78	-0.93	0.55	-2.16	3.30		-0.34
C$_{10}$	-0.88	0.03	0.29	-0.56	-1.81	-0.52	0.02	-2.31	0.50		-2.36
C$_{11}$	-0.36	-0.01	0.12	-0.25	-1.03	-0.25	0.14	-1.14	0.00		-1.38
Polynuclears >200 °C									-0.03	-3.19	-0.03
Total	-2.08	0.19	2.76	0.87	-4.50	-1.86	1.30	-5.06	7.35	-3.19	-3.19
8% 550 SiO$_2$/Al$_2$O$_3$ ZSM-5											
C$_7$	-0.03	-0.06	-0.07	-0.16	0.00	0.00	0.00	0.00	-0.04		-0.19
C$_8$	0.14	-0.02	-0.46	-0.34	0.10	-0.24	0.13	-0.01	2.44		2.08
C$_9$	-0.99	0.05	-0.38	-1.32	-1.38	-0.89	0.15	-2.12	3.12		-0.33
C$_{10}$	-0.08	0.04	0.12	0.08	-1.47	-0.57	-0.10	-2.14	3.52		1.48
C$_{11}$	0.22	-0.05	0.16	0.33	-1.60	-0.43	0.65	-1.38	0.00		-1.05
Polynuclears >200 °C									0.05	-2.04	0.05
Total	-0.74	-0.04	-0.63	-1.41	-4.35	-2.13	0.83	-5.65	9.09	-2.04	-2.04
8% 850 SiO$_2$/Al$_2$O$_3$ ZSM-5											
C$_7$	0.08	0.00	0.02	0.10	0.00	0.00	0.00	0.00	0.01		0.12
C$_8$	0.33	-0.02	-0.29	0.02	0.07	-0.15	0.14	0.06	0.83		0.90
C$_9$	-0.61	0.02	-0.05	-0.64	-0.69	-0.77	0.40	-1.06	1.14		-0.57
C$_{10}$	0.02	0.02	0.13	0.17	-0.89	-0.44	0.03	-1.30	0.78		-0.35
C$_{11}$	0.09	0.01	0.12	0.22	-0.52	-0.26	0.02	-0.76	0.00		-0.52
Polynuclears >200 °C									0.03	0.39	0.03
Total	-0.09	0.03	-0.07	-0.13	-2.03	-1.62	0.59	-3.06	2.79	0.39	0.39

indicated by the (C) in the Figure labels. The analysis of the light gasoline was done with the SE30 technique and the heavy gasoline by PIONA. Octanes were determined using a mini-engine technique. All yields, product properties, and compositions were adjusted to constant conversion using plots generated from runs at different cat-to-oil ratios. Conversion was effected little by the addition of ZSM-5 so these adjustments were minimal.

Yield Shifts

Yield shifts observed in the pilot plant are typical of those observed commercially (See Figure 1). Little, if any C_{2-} is produced by the ZSM-5 additives. The most predominant yield shifts are the increase in LPG and the decrease in gasoline yield. For the ZSM-5 additive with the 40 silica-to-alumina ratio, there appears to be a small increase in ethylene production and a small decrease in LCO while for the higher ratio additives, these yields appear unaffected. As the silica-to-alumina ratio of the additive increases, the absolute values of the incremental yield shifts of all of the products decrease.

The incremental LPG produced is very olefinic (See Figure 2). For the 40 silica-to-alumina additive, more propylene is produced than butylene on a weight basis which means substantially more on a molar basis. Little propane or

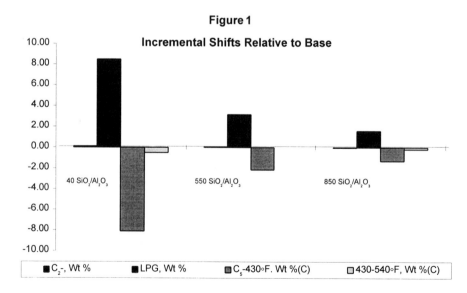

Figure 1

Incremental Shifts Relative to Base

Figure 2

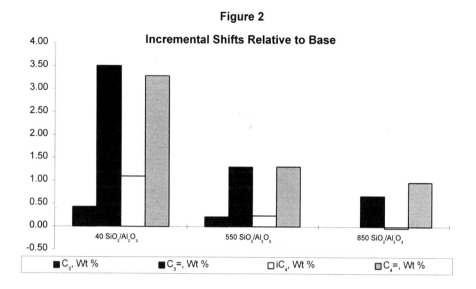

Incremental Shifts Relative to Base

Legend: ■ C_3, Wt % ■ C_3=, Wt % □ iC_4, Wt % □ C_4=, Wt %

X-axis categories: 40 SiO_2/Al_2O_3, 550 SiO_2/Al_2O_3, 850 SiO_2/Al_2O_3

n-butane is produced. Incremental isobutane yield is only a third of the butylene increase. As the silica-to-alumina ratio of the additive increases, the absolute values of these yield shifts decreases, although it is interesting to note that the production of propylene drops more rapidly than the production of butylenes. The selectivity shifts will be addressed later.

In Figure 3, shifts in lumped C_3's and C_4's along with the gasoline cuts are presented. As with the olefins, more total C_3's than C_4's are produced on a volume basis. This is true for all of the additives. The 40 SiO_2/Al_2O_3 additive produced a slightly bigger loss in light gasoline (LCN) than heavy gasoline (HCN). As the silica-to-alumina ratio is increased, it appears that HCN is preferentially lost relative to LCN. In fact, very little LCN is lost for the higher ratio additives.

Octane Shifts

Octane shifts are shown in Figure 4. For the 40 silica-to-alumina ratio additive, LCN MON increase is only about 20 percent of LCN RON increase. Octane increase in the HCN is greater as is the relative increase of MON to RON. At the higher silica-to-alumina ratios, all of the octane shifts decreased with RON decreasing more rapidly than MON.

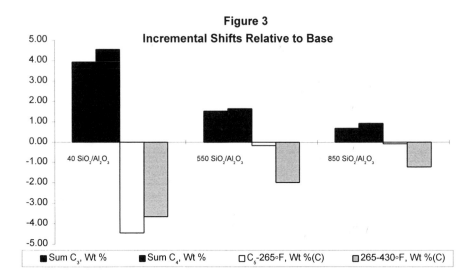

Figure 3

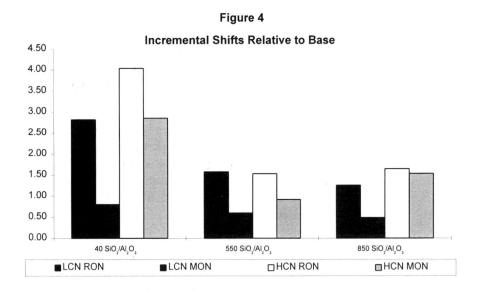

Figure 4

Relative Incremental Shifts

Although the magnitude of all of the shifts decreases with increasing silica-to-alumina ratio, their rate of change is different. Depending on refinery economics, these selectivity shifts may more than outweigh the cost of incremental additive required to achieve a given level of octane gain or olefin production.

In Figures 5 and 6, we show some yield and octane shifts relative to increase in C_4-. C_4- is chosen because it is a rough approximation of FCC wet gas rate. As seen in Figure 5, the fraction of propylene remains approximately constant while the fraction of butylene increases. This could be advantageous to a refiner with an alkylation unit. Figure 5 also illustrates that this stream becomes richer in isobutylene, an advantage to a refiner making MBTE. The last ratio in Figure 5 shows that the incremental loss in HCN relative to loss in total gasoline. For the 40 silica-to-alumina ratio additive the losses are fairly evenly balanced between light and heavy gasoline. For the higher ratio additives, the loss is essentially all HCN (90 %). This could be of advantage to a refiner making reformulated gasoline for which HCN is in general an undesirable blending component.

In Figure 6, octane-related ratios are presented. The first two ratios are a measure of octane gain for a given wet gas capacity (C_4-). It can be seen that as

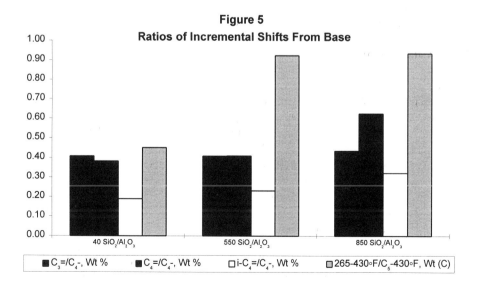

Figure 5
Ratios of Incremental Shifts From Base

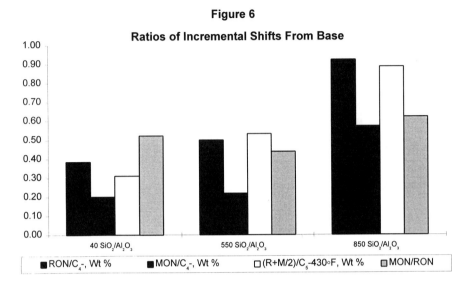

Figure 6

additive silica-to-alumina ratio increases, this measure increases for both RON and MON. Looking at it in a slightly different way, the third ratio shows that the amount of octane gain for a given gasoline loss increases steadily with increasing silica-to-alumina ratio. The last ratio indicates that there is a small increase in MON relative to that of RON. Although modest, this is significant in that sensitivity (RON - MON) generally increases with increasing octane. In this case, MON keeps up with this trend and even surpasses it.

Effect of Activity/Concentration on Selectivities

One concern in the selectivity changes observed is the different activities of the additives. If selectivities changed simply with the extent of reaction, conclusions reached above would be invalid. To address this question, data is presented on a 40 silica-to-alumina additive run at three different concentrations: 3%, 8%, and 15%. Figure 7 displays incremental selectivity parameters of interest. There is little change in the octane/C4- selectivity parameters, and the butylene/C_4- selectivity decreases only slightly at the very highest additive level. There is little change in the HCN/whole gasoline selectivity parameter with concentration. This is clearly differs from the observations made in interpreting the effect of silica-to-alumina ratio

It must be concluded from this data that comparison of the selectivity parameters at different extents of reaction is valid in this instance. This

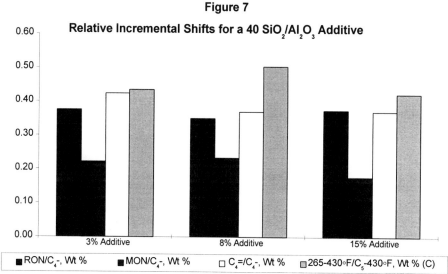

Figure 7

Relative Incremental Shifts for a 40 SiO$_2$/Al$_2$O$_3$ Additive

Legend: ■RON/C$_4$-, Wt %   ■MON/C$_4$-, Wt %   □ C$_4$=/C$_4$-, Wt %   ▨265-430∘F/C$_5$-430∘F, Wt % (C)

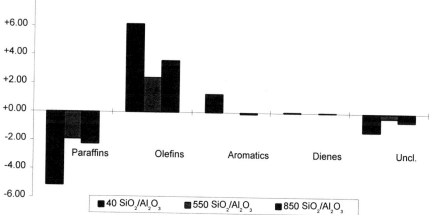

Figure 8

Shifts in Light Gasoline Composition (Vol %)

Legend: ■40 SiO$_2$/Al$_2$O$_3$ ■550 SiO$_2$/Al$_2$O$_3$ ■850 SiO$_2$/Al$_2$O$_3$

conclusion may not be valid at significantly different operating conditions or base octane.

Gasoline Compositional Shifts

Shifts in light gasoline composition are shown in Figures 8, 9, and 10. Figure 8 shows the shifts in the broad molecular categories. Consistent with general

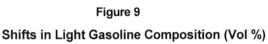

Figure 9

Shifts in Light Gasoline Composition (Vol %)

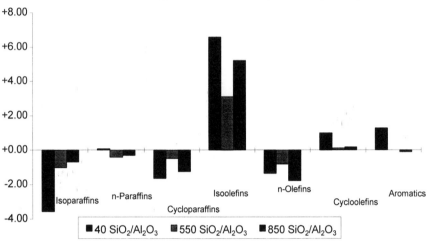

Figure 10

Shifts in Light Gasoline Composition (Vol %)

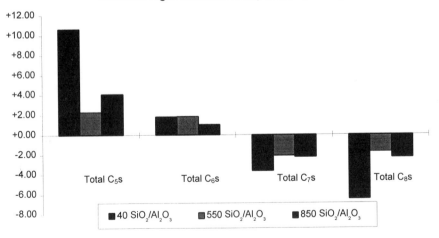

theory, paraffins are reduced and olefins increase. The reduction in paraffins comes mainly from the C_7 and C_8 range, while the increase in olefins occurs in the C_5s. There is a slight increase in aromatics for the 40 silica-to-alumina ratio due to concentration. As the silica-to-alumina ratio increases, the absolute values of the shifts are reduced and there is no apparent increase in aromatics.

In Figure 9, the effects on branching are displayed. Paraffin reduction occurs for all types of paraffins, surprisingly for isoparaffins. Isoolefins show the biggest increase. As the silica-to-alumina ratio of the additives increase, paraffin reduction is substantially less, while isoolefin increase does not suffer as much.

Figure 10 shows the shifts between carbon numbers in the light gasoline. The net shifts are from C_7s and C_8s to C_5s and C_6s, primarily C_5s. Increasing silica-to-alumina ratio seems to simply reduce the overall magnitude of these shifts.

Shifts in heavy gasoline composition are shown in Figures 11, 12, and 13. Figure 11 shows the shifts in broad molecular categories. In this case, saturates are affected very little. Aromatics increase due to concentration, and olefins are reduced. As silica-to-alumina ratio is increased, the increase in aromatics concentrations diminished substantially as is the magnitude of the olefin reduction.

In Figure 12, we see that the olefin reduction is more in the isoolefins than the normal olefins. Increasing the silica-to-alumina ratio diminishes the magnitude of the isoolefin reduction, but does not seem to affect the normal olefin reduction.

Finally, in Figure 13, we see the shifts among carbon numbers. C_8s are increased substantially by the 40 silica-to-alumina ratio additive. It is

Figure 11

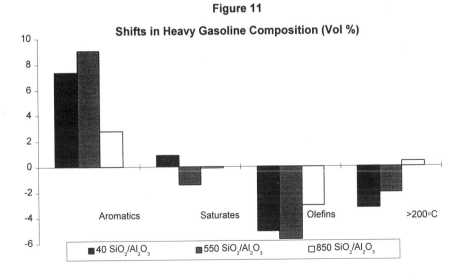

Figure 12

Shifts in Heavy Gasoline Composition (Vol %)

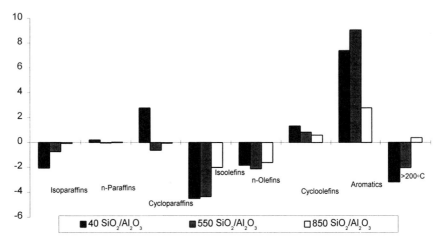

Figure 13

Molar Shifts in Heavy Gasoline Composition

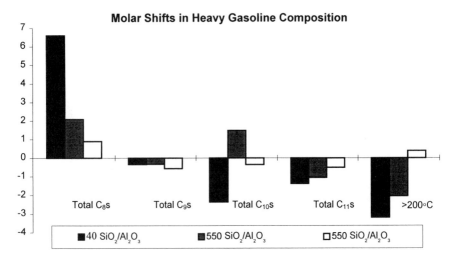

interesting to note that C_8s were reduced about the same amount in the light gasoline. Thus, C_8s seem to be unaffected on the whole. Shifts within this boiling range seem to be small no matter what the carbon number suggesting that gasoline reduction is uniform with carbon number. Increasing silica-to-alumina ratio even further minimizes these shifts.

DISCUSSION

The results from the 40 silica-to-alumina zeolite are typical for commonly used commercial ZSM-5 additives quite consistent with the presently accepted mechanism as detailed earlier in this paper. Significant amounts of propylene and butylenes are formed with propylene predominating. The concentration of C_5s in the light gasoline is increased dramatically. Aromatics are increased in both the light and heavy gasoline through concentration. Paraffins and straight chain olefins are reduced, and there is a clear increase in branched olefins, particularly in the light gasoline. What happens as the silica-to-alumina ratio of the ZSM-5 crystal is increased? This topic has been discussed in previous publications[6,7], and will be expanded on here in the context of the reported data.

Abbot and Wojciechowski[8] demonstrated that the significant olefin cracking reactions that would take place with a ZSM-5 FCC additive would be monomolecular. These would be associated with the alumina acid site on the zeolite. As the silica-to-alumina ratio of the ZSM-5 crystal is increased, the number of cracking sites associated with the alumina decline. Thus, the cracking reaction would decline. Isomerization reaction rate will also decline, but at a lower rate[6]. Hydrogen transfer, a bimolecular reaction, would decline even more rapidly. However, this would not be expected to be significant because this reaction is much slower for the ZSM-5 catalyst relative to the base faujasitic catalyst[9].

Thus, increasing the silica-to-alumina ratio of the additive crystal should reduce the cracking reactions. The results of this study are clearly consistent with this as evidenced from the reduced yields of C_3-C_5 olefins. The contribution of the C_5 olefins in the light gasoline to octane is clearly diminished as silica-to-alumina ratio is increased.

Reduction of the cracking reaction relative to the isomerization reaction also allows more opportunity for methyl isomers formed by the isomerization of a linear olefin to form an equilibrium distribution rather than the 2-position as favored by the shape selectivity of ZSM-5[10]. Cracking of the isomer with the methyl in the two position preferentially forms propylene. Thus, shifting the position of the methyl group away from the 2-position will yield more butylenes relative to propylene as seen in the pilot plant results. Because cracking increases more rapidly with carbon number than isomerization[8], it is only for the higher carbon numbers (8 and 9) that this becomes a significant phenomena.

The changes in the relative disappearance of light and heavy gasoline as the silica-to-alumina ratio increased is also explained by the dependence of the cracking rate on carbon number. The olefins in the light gasoline are more refractory and require the higher cracking activity of the 40 silica-to-alumina ratio ZSM-5 additive to crack. The higher molecular weight olefins in the heavy gasoline crack more easily thus responding more readily to the higher ratio additives. The lower cracking activity of these additives also could explain the absence of the small ethylene increase and LCO decrease observed with the lower ratio material.

As mentioned above, octanes in the light gasoline are reduced due to less C_5 olefins resulting from cracking of C_{8+}. However, since there is little light gasoline loss, the octane gain/gasoline loss ratio improves. There is no aromatics concentration effect with the higher ratio additives resulting in a slight octane debit.

The heavy gasoline octanes are a more complicated situation. In this case, concentration of aromatics should significantly affect octane. Because cracking is reduced as the silica-to-alumina ratio is increased, less octane gain due to aromatics is observed. On the other hand, there is less gasoline loss. It is clear from the data that both branched and linear olefin concentrations are reduced despite the concentration effect. These have been cracked into light olefins. In the case of the branched olefins, these represent an octane debit. However, it should be noted that the loss of isoolefins is lower for the 850 silica-to-alumina ratio zeolite. This could be accounted for by both reduced cracking or increased isomerization.

Finally, it would be expected that with isomerization playing a larger role, that MON would be more favored. This would result from the large MON contribution of branched compounds. As noted in Figure 6, the incremental MON increase is about 50% of the RON increase for the 40 silica-to-alumina additive rising to approximately 60% for the 850 silica-to-alumina additive. This should be contrasted to a value of about 30% if the octane increase were obtained by reactor temperature or base catalyst.

CONCLUSIONS

As the silica-to-alumina ratio of the ZSM-5 crystal of an FCC octane additive is increased, the incremental ratios of octane/LPG, octane/gasoline loss, $C_4=/C_3=$, and MON/RON increase. Less relative light gasoline loss is experienced as the silica-to-alumina ratio is increased. These phenomena are consistent with a mechanism in which olefin intermediate cracking is reduced relative to isomerization.

REFERENCES

[1] Miller, S. J., and Bishop, K. C., III, U. S. Patent 4,340,465 (1982).

[2] Miller, S. J., U. S. Patent 4,309,276 (1982).

[3] Buchanan, J. S., Applied Catalysis, 74 (1991) 83-94.

[4] Madon, R. J., J. Catal., 129 (1991) 275-287.

[5] Anderson, C. D., Dwyer, F. G., Koch, G., and Niiranen, P., Proc. Ninth Iberoamerican Symposium on Catalysis, Lisbon (1984) 247.

[6] Miller, S. J. and Hsieh, C. R., in Fluid Catalytic Cracking II: Concepts in Catalyst Design; Occelli, M. L., Ed.; ACS Symp. Series 452; American Chemical Society (1991) 96-108.

[7] Miller, S. J., Hsieh, C. R., Kuehler, C. W., Krishna, A. S., NPRA Annual Meeting Paper AM-94-58.

[8] Abbot, J., and Wojciechowski, B. W., Can. J. Chem. Eng., 63 (1985) 462-469.

[9] Miller, S. J., Stud. Surf. Sci. Catal., 38 (1987) 187.

[10] Jacobs, P. A., Martens, J. A., Weitkamp, J., and Beyer, H. K., Faraday Discussions of the Chemical Society, 72 (1982) 353-368.

STRUCTURE/ACTIVITY CORRELATIONS ON NICKEL CONTAMINATED FLUID CRACKING CATALYSTS

Geoffrey L. Woolery, Maria D. Farnos, Augusto R. Quinones* and Artie Chin
Mobil Technology Company, Paulsboro Technical Center
Paulsboro, NJ USA 08066

I. <u>Abstract</u>

X-ray Absorption Spectroscopy and fixed fluid bed pilot plant testing were used to characterize the structural and chemical properties of Ni contaminated FCC catalysts after commercial and laboratory-simulated equilibration. Commercial FCC equilibrium catalysts, obtained from the regenerator, contain both dispersed nickel oxide and nickel aluminate; the relative concentration of these species varies based on catalyst type and unit operating conditions. Laboratory-simulated equilibrium catalysts, metalated to 2000-3000 ppm Ni, also contain nickel oxide and nickel aluminate. The relative concentration of nickel aluminate increases with increasing

*Currrent address: Akzo-Nobel Chemicals, Research Center, Pasadena TX

deactivation severity. A series of different catalyst technologies were examined where the primary difference in catalyst was matrix type.

Ni aluminate formation, relative to Ni oxide, is found to vary significantly depending on catalyst type. Ni-promoted dehydrogenation activity decreases with decreasing nickel oxide concentration. These results suggest that catalyst formulations that promote Ni aluminate formation relative to Ni oxide will show reduced metal-promoted dehydrogenation activity and improved FCC performance.

II. Introduction

It is well known that metal contaminants, principally nickel, promote dehydrogenation reactions in the FCC riser (Cimbalo et al., 1972; Campagna et al., 1983). Their effect on FCC operations and performance is considered when evaluating catalyst management strategies for processing resid and other high metal-containing feeds. Knowledge of the rate of metals build-up as well as the absolute metals concentration is critical since freshly deposited metals are more active than metals that have been equilibrated through numerous cracking/regeneration cycles (Mitchell, 1980). In order to investigate the chemical and structural changes responsible for these activity differences, a series of Ni-containing FCC catalysts was studied by X-ray Absorption Spectroscopy (XAS). Previous studies have shown XAS to be a valuable tool for determining the state of vanadium on FCC catalysts (Woolery et al., 1988; Sajkowski et al., 1989). Commercial E-cats were compared with laboratory deactivated catalysts employing a number of different silica sol and alumina sol matrix technologies. Structural properties were then compared with catalyst selectivities as determined by pilot scale fixed fluid bed testing. Passivation chemistry was studied by treating equilibrium catalysts with reducing gases (such as H_2), which has been claimed to suppress metal promoted dehydrogenation reactions (Davis and Rase, 1986; Lee, 1989).

III. Experimental

A. Catalysts:

Regenerated equilibrium catalysts were obtained from several full combustion FCCU regenerators. Coke on catalyst ranged from 0.05 to 0.10 wt%. Laboratory catalysts were prepared by impregnation with Ni naphthenate to 2000-3000 ppm Ni, followed by nitrogen calcination at 1000°F for 2 hr. to remove organic and air calcination at 1200°F for 3 hr. Except where noted, the catalysts were then steam deactivated for 5 to 20 hr. at 1400°F and 35 psig in a continuous steam atmosphere which cycled between air and 5% propylene in nitrogen (with nitrogen purge, i.e. CPS procedure), in a modified procedure to that reported previously (Cheng et al., 1992). A series of different REUSY commercially manufactured catalysts was examined whose primary difference was the type of matrix technology employed. Key catalyst chemical and

physical properties are given in Table 1.

Table 1: Catalyst Properties

	Cat A	Cat A	Cat B	Cat C
zeolite type	REY	REY	REUSY	REUSY
ZSA, m^2/g	70	85	120	110
matrix type	Al sol	Al sol	Si sol + Al	Si sol + Al
MSA, m2/g	25	25	40	35
RE_2O_3, wt%	3.5	3.5	2.0	2.7
Ni, ppm	3300	4900	1000	2000
V, ppm	980	74	2000	2200
Na, ppm	9400	4500	5000	5100
C, wt%	0.05	<0.05	0.10	0.05
deactivation	E-cat	lab cat	E-cat	E-cat

	Cat D	Cat E	Cat F	CatG
zeolite type	REUSY	REUSY	REUSY	REUSY
ZSA, m^2/g	155	145	165	100
matrix type	Si sol + Al	Si sol + Al	act. clay	Si sol + Al
MSA, m^2/g	30	45	75	35
RE_2O_3, wt%	1.6	1.6	1.6	0.9
Ni, ppm	2500	2500	2500	3000
V, ppm	30	30	30	5000
Na, ppm	4000	000	4000	8000
C, wt%	<0.05	<0.05	<0.05	<0.05
deactivation	lab cat	lab cat	lab cat	Ecat

B. <u>XAS Data Collection and Analysis</u>:

Nickel K-edge EXAFS and near edge data were obtained on beamline

X-11A at the National Synchrotron Light Source. Catalysts were examined in

fluorescence mode at room temperature. In-situ H_2 treatments on equilibrium

catalysts were performed in flowing gas at 950°F, 0 psig for durations of 1 and

30 minutes. Ni reference compounds were diluted with boron nitride

and examined in transmission mode. Ni foil was used as the metallic

reference, $Ni(NO_3)_2 \cdot 6H_2O$ was used as the Ni-O reference and NiS

was used as a Ni-Al(Si) reference. Ni aluminate was prepared according to

published procedures (Sand, 1982) and the other references were obtained from

commercial sources. The structural integrity of all materials used as references

was verified by X-ray diffraction. I_0, I_f and I photon intensities were

monitored using ionization chambers. Third order harmonics were essentially

eliminated by detuning the incident photon energy from the Si(111) crystal

monochromator by ~20%. The X-ray storage ring was operating at an electron

energy of 2.5 GeV at currents of 75-200 mA. Data reduction and analysis were

performed as previously described (Lee et al., 1981) using k_2 and k_3 weighted

EXAFS modulations transformed over a range of 2.5-12.5 A^{-1}. [Data were

limited to 12.5 A^{-1} due to trace levels of Cu on commercial FCC catalysts.]

Structural parameters were obtained by fitting to experimentally determined

phase and amplitude files of the aforementioned reference compounds.

C. FFB Testing:

Fixed fluid bed testing was performed as previously described (Sapre and Leib, 1990). The feed properties are given in Table 2. Reactor temperature was maintained at 850°F during the 5 minute run; feed rate was 18g/min. and catalyst loading was 180g. Coke was analyzed with a LECO

Table 2: Fixed Fluid Bed Feed Properties

Light East Texas Gas Oil

API Gravity	36.3
Aniline Point, °F	168
Sulfur, wt%	.128
Total nitrogen, wt%	<.03
Basic nitrogen, ppm	33
CCR, wt%	.01
Hydrogen Content, wt%	13.33
Molecular Weight	223
Metals	
Ni	<.1
V	<.1
Cu	.1
Fe	.8
Distillation, °F	
IBP	460
50%	546
90%	650

carbon analyzer, gases were analyzed by on-line GC, C_1-C_6 composition in gasoline was analyzed by GC (ASTM M-1363) and liquid product was analyzed by simulated distillation (ASTM D2887). Conversion is defined as coke + 465°F- product. Coke selectivity, Kc, is defined as coke on catalyst/kinetic conversion [vol% conversion/(100 - vol% conversion)]. H-factor is defined as 100* moles of H_2/(moles of C_1+C_2). This FFB test has been used by Mobil for over 25 years to monitor commercial equilibrium catalyst performance and is found to correlate very well with commercial activities and selectivities.

IV. Results and Discussion

X-ray Absorption spectra were obtained on a commercial REY catalyst (Cat A) and the same base catalyst after Mitchell impregnation and mild laboratory steaming for 5 hr., 15 psig air/steam mix (50/50) at 1400°F. A comparison of the background corrected, k^3 weighted EXAFS modulations are shown in Figure 1. The Fourier transformations (Ft) of the EXAFS modulations for these two catalysts are compared in Figure 2. A significant difference between the short range structure surrounding Ni is evident from these comparisons. Analysis of the laboratory catalyst, deactivated under mild steaming conditions to simulate "fresh" metals, shows that the Ni environment is similar to a dispersed Ni oxide; the scattering is dominated by Ni-O first

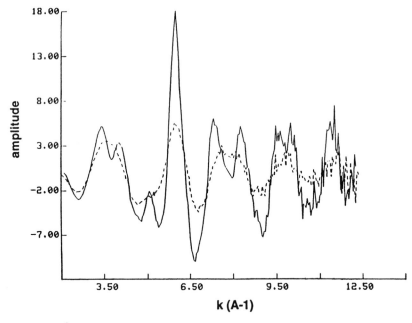

Figure 1. k^3 weighted EXAFS modulations for equilibrium (solid) and laboratory (dashed) Cat A

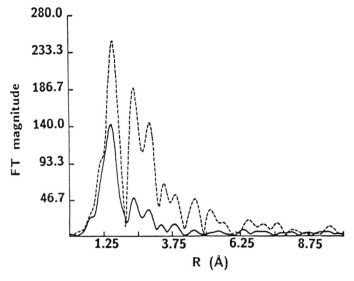

Figure 2. EXAFS Fourier transformations for equilibrium (dashed) and laboratory (solid) Cat A

shell contributions at about 2.0Å. In addition to a first shell Ni-O scatterer, the

commercial E-Cat contains strong outershell scattering contributions. The Ft

of Cat A Ecat is compared with that of neat Ni aluminate in Figure 3. It can be

seen from this comparison that the outershell scattering of the Ni aluminate is

very similar to that of the E-cat. To better characterize the Ni structure on Cat

A Ecat, the outershells of the Ft were backtransformed into k space and fitted to

amplitude and phase files from the neat aluminate (Lee, 1981). The Ecat

scattering could not be adequately modeled by fitting to the neat aluminate

alone; an additional scattering contribution at about 3.0 Å was required . The

best fit was obtained when this additional scatter was Ni and this seemed

chemically reasonable since the Ni-Ni distance in Ni oxide is 3.0 Å The

backtransformed outershells of the Ecat are compared with the model fit in

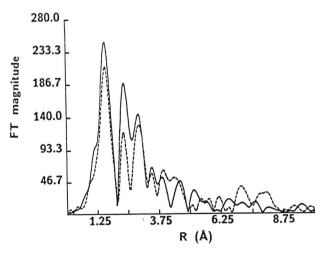

Figure 3. EXAFS Fourier transformations of Cat A (solid) and NiAl$_2$O$_4$ (dashed)

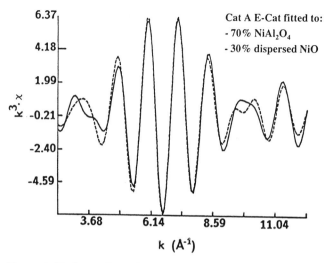

Figure 4. Back transformed outer shell doublet of Cat A fit to NiAl$_2$O$_4$ and NiO

Figure 4. For this comparison the Ecat was fit to 70% Ni aluminate and 30%

of a dispersed Ni oxide (i.e a Ni oxide with an average particle size of 10-15Å,

or a particle with an average Ni-Ni coordination number of 6.0, compared with

12.0 for bulk Ni oxide).

Due to the differences in total metal content and overall catalyst

activity, no attempt was made to draw conclusions from the pilot plant

selectivity differences between Cat A Ecat and the mildly deactivated lab cat.

These results did however indicate that lab deactivation, under these mild

conditions (lab catalyst conversion was 15 vol% higher than E-cat), leads to a

very different Ni species than observed on commercial equilibrium catalyst and

most likely represents the Ni structure of "fresh" metals. To see if passivation

via reducing gas had any effect on Ecat Ni structure for this catalyst system we

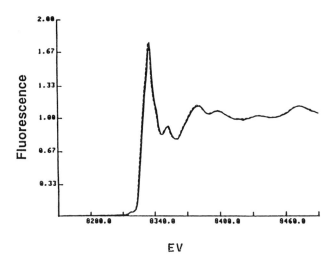

Figure 5. Ni K edge spectra of Cat A, as received (solid), and after in-situ reduction (dashed) at 950°F for 30 min. in flowing H_2

performed n-situ treatment of the REY E-cat with H_2 for 30 minutes at 950°F.

This treatment had no effect on the Ni near edge spectra, as shown in Figure 5, suggesting no reduction or structural change of the Ni species. It is not surprising that the Ni aluminate is unreactive under these conditions (Tatterson and Mieville, 1988) but the inability to reduce the Ni oxide component suggests that it may be encapsulated within the Ni aluminate.

Once the presence of Ni aluminate was determined on REY E-cat, several additional commercial REUSY E-cats were examined. Figure 6 compares the near edge spectra (after linear background removal) of two different REUSY catalysts (Cat B and Cat C) from two different full combustion units. Qualitatively, Cat B contains more Ni as the aluminate than

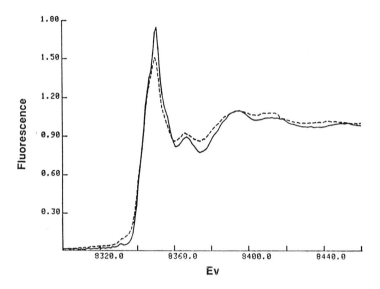

Figure 6. Ni K edge spectra of Cat B (solid) and Cat C (dashed) from full burn FCCUs

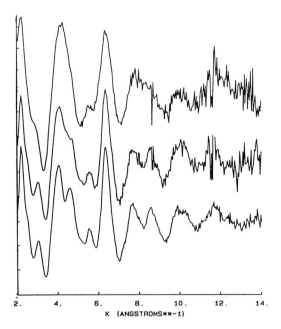

Figure 7. k^2 weighted EXAFS modulations comparing Lab Cats D-5 (bottom), E-5 (middle), and F-5 (top)

Cat C. This determination was made based on Ni K-edge data from our

reference compounds. However, it is not clear whether these differences are

due to catalyst type (of which the primary difference is matrix type) or aging

differences due to different unit operating conditions.

In order to examine the effect of catalyst formulation on Ni

oxide/aluminate formation, a series of REUSY catalysts with comparable

zeolite and rare earth content, but different matrix technologies was examined.

The background corrected k^2 weighted EXAFS modulations for Catalysts D, E,

and F are shown in Figure 7, after CPS steam deactivation for 5 hr. Fixed fluid

bed performance data are shown in Table 3. The Ni aluminate to Ni oxide ratio

increases from D > E > F, as evidenced by the increasing intensity of a doublet

appearing in the EXAFS modulations at 3.1, 4.7, 5.7, and 8.0 A^{-1}. (This

doublet is due to interference of the first shell Ni-O scattering and outershell

Ni-Ni and Ni-Al scattering, from the aluminate). The FFB results show that, as

Ni aluminate content increases relative to the oxide, H_2 make and coke make

both decrease. The XAS data indicate that different types of FCC matrix

technologies promote the formation of Ni aluminate over Ni oxide to different

extents. The FFB data demonstrate that these structural differences lead to

differences in Ni promoted dehydrogenation activity; those catalysts which

promote Ni aluminate formation show superior Ni tolerance. The k^2 weighted

EXAFS modulations for Catalyst E, after both 5 and 20 hr. CPS steamings are

Table 3: Fixed Fluid Bed Results

	Cat A	Cat A	Cat B	Cat C
deactivation	E-cat	5 hr lab	E-cat	E-cat
conversion, vol%	60	75	64	60
gasoline, vol%	53	56	51	47
Kc	0.67	0.88	0.25	0.34
H-factor	NA	NA	200	400

	Cat D	Cat E	Cat E	Cat F
deactivation	5 hr lab	5 hr lab	20 hr lab	5 hr lab
conversion, vol%	74	75	69	73
gasoline, vol%	57	56	51	55
Kc	0.30	0.37	0.26	0.40
H-factor	150	325	300	375

where   Kc = coke on catalyst/Cr

Cr = vol% conv./(100 - vol% conv.)

H-factor = 100 x [moles H2/(moles of C1 + C2)]

shown in Figure 8. As compared to the 5 hr. sample, the 20 hr. sample

contains only marginally more Ni aluminate, indicating that this matrix

technology promotes Ni aluminate formation relatively quickly, and certainly to

a muck greater extent than Catalyst A. The FFB data show a slightly lower H_2

yield for the more aged Ni sample, Cat E-20. Note that coke factors (Kc) can

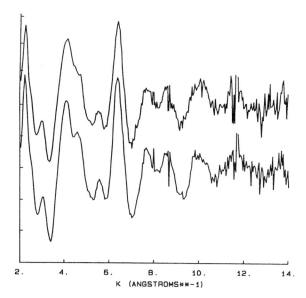

K (ANGSTROMS**-1)

Figure 8. k^2 weighted EXAFS modulations of Lab Cats E-5 (top) and E-20 (bottom)

not be compared directly here since the 5 hr. sample has a much greater unit

cell size (and thus higher coke make) than the 20 hr sample.

As another example of how H_2 and coke make in an FFB evaluation

can be related to Ni aluminate content (relative to dispersed Ni oxide) we

compared a commercial Ecat to lab catalyst deactivated by cyclic metals

laydown (i.e. aged over time in a pilot unit) to the same catalyst deactivated by

the conventional Mitchell method. To roughly quantify the degree of Ni

aluminate present in the samples, the integrated Ft area of the outershells was

calculated. As the outershell area increases, the relative concentration of Ni

aluminate increases. The inverse of the magnitude was plotted vs. the coke

and H_2 make from the FFB data. The results are shown in Figure 9 and 10.

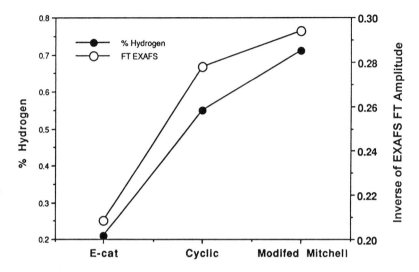

Figure 9. Correlation between FFB constant conversion H₂ yields and EXAFS outershell Ft amplitude

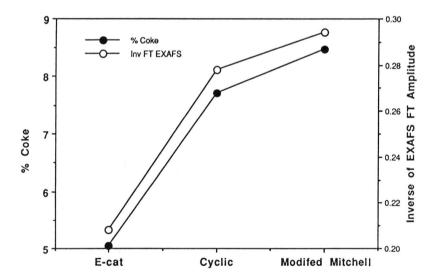

Figure 10. Correlation between FFB constant conversion coke yields and EXAFS outer shell Ft amplitude

The H_2 and coke yields track surprising well with the aluminate content

determined in this admittedly pedestrian fashion.

A final set of experiments was performed on an Ecat taken from a

resid FCC unit, Cat G. This Ecat had a very high concentration on Ni oxide,

due in part to the matrix type but probably also form the high catalyst turnover

rat , and therefore low average metals age, in the unit. After collecting Ni K

edge data on the as received Ecat, we treated the sample in flowing H_2 at 950°F

for 30 min., as was done on Cat A. In contrast to what we observed on that

sample, the resid Ecat showed a significant change in the edge features,

indicating reduction of the Ni. We then repeated the H_2 treatment on the fresh

Ecat, but this time for only 1 min. Again a similar change in edge structure

was observed indicating the this Ni species is highly reactive. These data are

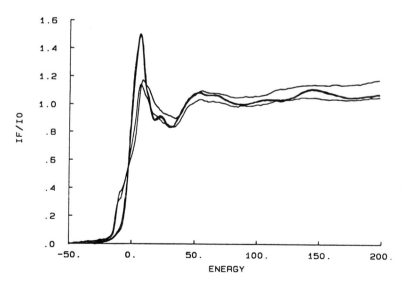

Figure 11. Ni K edge spectra for Cat G, as recieved (heavy), after 30 min. in-situ reduction (top) , and 1 min. in-situ reduction (bottom) at 950F in flowing H_2

summarized in Figure 11. These results taken together with our results on Cat

A suggest that the effectiveness of reducing gas passivation of Ni containing

catalysts may be very dependent on the Ni structure, and therefore the type of

catalyst used.

V. Conclusions

Ni on metal contaminated FCC catalysts exits in two primary

structures; dispersed Ni oxide and Ni aluminate. The oxide structure is favored

initially as Ni is deposited on catalyst but aluminate formation occurs with

catalyst aging. The aluminate is less active than the oxide for dehydrogenation,

possibly due to the inertness of the aluminate. Different matrix types promote

Ni aluminate formation, relative to the oxide, to different degrees. FFB testing

suggests that those catalysts promoting formation of the aluminate will exhibit

superior metals tolerance in FCC. These types of characterization results can

be used to supplement pilot plant and bench scale testing to help rank the

metals tolerance of Ni containing FCC catalysts.

VI. Acknowledgments

The authors would like to acknowledge Mobil Technology Company

for permission to publish this work . The XAS portion of this work was

conducted on Beamline X-11 at NSLS and is supported by the Division of

Materials Science of the US Department of Energy under contract No. DE-AS05 80ER10742. The authors also wish to thank Drs. Art Chester and Tony Fung for helpful discussions.

VII. References

1. R. N. Cimbalo, R. L. Foster, S. J. Wachtel, (May 15, 1972). Oil and Gas Journal, 112-122.

2 R. J. Campagna, A. S. Krishna, S. J. Yankin, (Oct. 31, 1983). Oil and Gas Journal, 128-134.

3. A. N. Mitchell, (1980). I&EC, Prod R&D, 19, 209-213.

4. G. L. Woolery, A. A. Chin, G. W. Kirker, and A. Huss, Jr., (1988). "Fluid Catalytic Cracking: Role in Modern Refining", ACS Symposium Series 375, Mario L. Occelli (ed.) 215-228

5. D. J. Sajkowski, S. A. Roth, L. E. Iton, B. L. Meyers, C. L. Marshall, T. H. Fleisch and W. N. Delgass, (1989). Appl. Catal., 51(2), 255-262.

6. T. R. Davis and H. F. Rase, (1986). I&EC, Fund. 25, 581-588.

7. F. M. Lee, (1989). Ind. Eng. Chem. Res. 28(7), 920-925, and references within.

8. W.-C. Cheng, M. V. Juskelis, and W. Suarez, (1992). AIChE Annual Mtg, Miami Beach, FL.

9. M. L. Sand, PhD Dissertation, (1982). Univ. of Delaware, Dept. of Chem Eng.

10. P. A. Lee, P. A. Citrin, P. M. Eisenberger, B. M. Kincaid, (1981). Rev. Mod. Phys. 53769.

11. A. V. Sapre and T. M. Leib, (1990). "Fluid Catalytic Cracking II", ACS Symposium Series 452, Mario L. Occelli (ed.) 144-164.

12. D. F. Tatterson and R. L. Mieville, (1988). I&EC Res., 27, 1595-1599.

THE EFFECT OF THE OXIDATION STATE OF VANADIUM ON THE SELECTIVITY OF FLUID CATALYTIC CRACKING CATALYSTS

John Allen Rudesill and Alan W. Peters
GRACE Davison
Division of W. R. Grace & Co.-Conn.
7500 Grace Drive, Columbia, MD 21044
(410) 531-4036

Experimental results with FCCU equilibrium catalysts show a strong effect of vanadium oxidation state on catalyst selectivity. Catalysts containing vanadium in the reduced +4 state as measured by TPR make much less coke and more gasoline than the same catalysts in the oxidized +5 state. The oxidation state of +5 can be achieved either in a full burn regeneration or by catalyst calcination, while equilibrium catalysts containing residual coke from a partial burn operation may contain vanadium in the reduced +4 state. These results have implications both for catalyst testing and for the operation of the FCCU in processing high vanadium residual oil.

I. Introduction

The evaluation of catalytic cracking catalysts involves a laboratory deactivation including a hydrothermal treatment at a high temperature, typically 760°C to 810°C (1), and an impregnation with nickel and vanadium, the metal contaminants usually found on equilibrium catalysts from an operating FCCU processing residual oil (2, 3). The effect of nickel and vanadium is to decrease activity and to change selectivity in the direction of increased amounts of coke and gas, especially hydrogen and methane (2-4). These effects are simulated by impregnating the catalyst with an appropriate amount of nickel and vanadium and steam deactivating (3-5), most effectively under cyclic conditions (6-9).

There remains the issue of testing after deactivation. After either unit deactivation or cyclic deactivation procedures, the catalyst and the metals on it will have had some definite exposure to a final oxidizing or reducing environment before activity and selectivity testing. The oxidation state of vanadium especially can have a significant effect on catalytic selectivity for coke and hydrogen (8,10). A measure of the oxidizing severity is the amount of carbon present on the regenerated catalyst. We have found that the oxidizing environment of the catalyst during deactivation affects the oxidation state of the metal as measured by TPR (Temperature Programmed Reduction) experiments and has a major effect on the observed selectivies of the catalyst being evaluated. Consequently, the final oxidizing environment must be as carefully controlled as the rest of the deactivation procedure of the test catalyst.

II. Experimental
In one set of experiments we selected equilibrium catalysts with a range of 0.03 wt.% to 0.39 wt.% carbon on catalyst from a number of commercial units. Since the amount of carbon on the regenerated catalyst is a measure of oxidizing severity in the regenerator, these catalysts represent a range of regenerator oxidation conditions. The catalysts were microactivity tested using a modified version of ASTM Test Method D 5154-91 (11). The modification consisted of the use of three receivers, one using water ice and two packed with dry ice. Reaction occurred at a reactor temperature of 528°C with a catalyst contact time of 30 seconds. Each catalyst was tested at three different catalyst to oil weight ratios, nominally 3,4, and 5. Values of the C/O, coke, and hydrogen at a particular conversion are obtained by interpolation. The properties of the feedstocks used in the MAT testing are given in Table 1. Each catalyst was tested both as received and after a two hour calcination at 705°C.

TPR analysis of these and other catalyst samples was carried out in a flowing 5% hydrogen stream. The sample (0.05 gms) was placed in a stream of flowing hydrogen and the temperature increased from 100°C to 800°C at a rate of 20°C/minute. Changes in hydrogen content are monitored by a thermal conductivity detector. A one electron reduction per hydrogen atom is assumed.

CO reduction was carried out in a vertical tube furnace at 733°C for two hours at 20,000 space velocity in a 5% CO / helium mixed gas.

Table 1. Gas oil feedstock properties.

API @ 60°F	22.5
Specific Gravity	0.9186
Aniline Pt., °F	163
Sulfur, Wt%	2.59
Total Nitrogen, Wt%	0.08
Basic Nitrogen, Wt%	0.03
Conradson Carbon, Wt%	0.25
Ni, ppm	0.8
V, ppm	0.6
Fe, ppm	0.6
Na, ppm	0.6
Refractive Index	1.51131
K factor	11.46
Simulated Distillation, °F	
10%	577
50%	750
90%	925
End	1038

Table 2. Analysis of Equilibrium Catalysts from Commercial FCC Units as Received (ASR) and Calcined.

Equilib Catalyst	A1	B1	C1	D1	D2	E1	E2
Wt% Carbon	0.03	0.07	0.08	0.16	0.17	0.31	0.39
ppm V	3325	1557	4424	4266	4172	5733	5829
μm V/gm	65	31	87	84	82	113	115
ppm Ni	1415	644	966	1994	1996	2833	2947
μm Ni/gm	24	11	16	34	34	48	50
ppm Sb	1100	226	2	481	482	710	1072
μmSb/gm	9.0	1.9	0.0	4.0	4.0	5.8	8.8
TPR, μm H/gm	211	185	194	122	131	143	131
After Calcination 2 Hours 705°C							
TPR, μm H/gm	191	222	255	234	238	266	285
Calc-ASR, μm H/gm	-20	37	61	112	107	123	154
Calc-ASR, μmH/μmV	-0.31	1.21	0.70	1.34	1.31	1.09	1.34

III. Results and Discussion

A. Equilibrium Catalysts from the Regenerator

Table 2 gives the compositions of seven equilibrium catalysts with moderate (2000 ppm Ni+V) to high (≥ 5000 ppm Ni+V) contaminant metal levels and with carbon on catalyst levels ranging from 0.03 wt% to 0.39 wt% carbon. The amount of reducible metal both before and after calcination was measured by TPR experiments. TPR curves for each of the equilibrium catalysts before and after calcination are shown in Figure 1. The peak at about 600°C has been identified as a vanadium reduction from +5 to +4 (12). Reduced vanadium on a spent FCCU catalyst has been independently identified as being

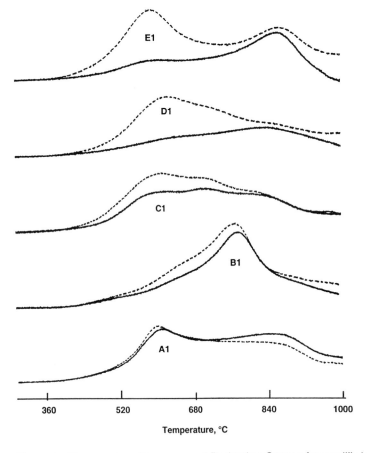

Figure 1. Temperature Programmed Reduction Curves for equilibrium catalysts (Table 2)both as received, solid line, and after calcination in air for two hours at 705°C, dotted line.

in the +4 state (13). The higher temperature peak at 680-700°C has been identified as Ni reduction from +2 to 0 (12), while the highest temperature peaks at $\geq$ 840°F are possibly inorganic iron, nickel or other inorganic silicates or aluminates. They do not play a role in the present work since they cannot undergo oxidation or reduction in the environment of the FCCU reactor. The occurrence of high temperature reducible species on the catalysts besides Ni or vanadium can be observed by comparing the amount of the total amount of reducible species after calcination with the calculated micro moles of hydrogen required to reduce nickel and vanadium, assuming reductions from V^{5+} to V^{4+} and from Ni^{2+} to Ni^0, Table 2. The total reducibility of the catalyst is greater than the sum of the micromoles of reducible nickel and vanadium.

The experiments described here depend on being able to measure differences in reducible species and relating these differences to the amounts and the oxidation states of vanadium and nickel. The catalyst samples with the least carbon labeled A1, B1, and C1 are from a more oxidizing environment, while the catalysts with higher amounts of carbon, D and E, are from a less oxidizing (partial burn) regenerator operation. The differences in the temperature programmed reduction curves, Figure 1, support this conjecture. The amount of hydrogen picked up by samples A1, B1, and C1 with less than 0.1% carbon on the as received equilibrium catalyst is nearly the same as the amount of hydrogen picked up after laboratory calcination in air. For sample A1 with 0.03 % carbon the TPR curves in the vanadium region are identical, while for the other low carbon samples B1 and C1 there appears to be a small amount of reduced vanadium species. The TPR curves of samples D and E with higher coke show almost complete reduction of the vanadium on the as received samples compared to the calcined samples. The quantitative comparisons, Table 2, of the amounts of reducible vanadium with the differences in the TPR curves support the assignment of +4 as the oxidation state of the reduced vanadium, with the possible occurrence of a small amount of vanadium in the +3 state.

The effect of calcination on the selectivity and activity of these catalysts as measured by the microactivity test is given in Table 3. Activity is measured by comparing the catalyst to oil ratio required to achieve a particular activity. A less active catalyst will require a higher catalyst to oil ratio. The activity of the calcined samples compared to the activity of the as received samples increases only for the samples D and E with the higher amount of carbon on the as received catalyst. For the samples with less carbon the calcined activities are less than the as received activities. This result implies that partially reduced catalysts are more active than calcined catalysts provided the contaminant coke is less than 0.1%.

Table 3. Comparisons of the activities and the coke selectivities of the equilibrium catalysts of Table 2 as-recieved and after calcination for two hours at 705°C.

Eq. Catalyst	A1	B1	C1	D1	D2	E1	E2
Wt. % Carbon	0.03	0.07	0.08	0.17	0.16	0.31	0.39
ppm V	3325	1557	4424	4172	4266	5733	5859
μmV/gm Catalyst	65	31	87	82	84	113	115

Activity Comparisons (Differences in catalyst /oil ratio required to achieve equivalent conversion as shown)

	A1	B1	C1	D1	D2	E1	E2
% Conversion	60	67	67	60	60	54	54
KC (conv/(1-conv))	1.50	2.03	2.03	1.50	1.50	1.17	1.17
Catalyst/Oil Ratio							
as received	3.8	3.9	4.7	4.3	3.9	4.5	4
Calcined 2 Hr. 1300F°	4.3	4.6	5	4.1	3.9	3.7	3.5
Activity Difference, C/O units							
Calcined - As Received	0.5	0.7	0.3	-0.2	0	-0.8	-0.5

Coke Selectivity Comparisons
As Recieved MAT

	A1	B1	C1	D1	D2	E1	E2
Wt. % Coke on Feed	3.9	4.3	4.7	3.8	3.3	4.9	3.9
Coke/KC	2.60	2.12	2.31	2.53	2.20	4.17	3.32
Delta Coke	1.03	1.10	1.00	0.88	0.85	1.09	0.98
Calcined 2 Hr. 1300F°C							
Wt. % Coke on Feed	4.6	4.8	5.4	4.8	4.2	5.9	6.4
Coke/KC	3.07	2.36	2.66	3.20	2.80	5.03	5.45
Delta Coke	1.07	1.04	1.08	1.17	1.08	1.59	1.83
Coke Selectivity Differences (Calcined - As Received)							
Coke/KC	0.47	0.25	0.34	0.67	0.60	0.85	2.13
Delta Coke	0.04	-0.06	0.08	0.29	0.23	0.51	0.85
μm Carbon/gm Catalyst	36.21	-49.24	66.67	239.18	192.31	421.42	711.31
μm Carbon/μm V	0.55	-1.61	0.77	2.92	2.29	3.74	6.18

Coke selectivities are dramatically affected by the vanadium oxidation state. The coke as a percent of feed measured at the same activity was in all cases significantly greater for the calcined samples. However in the case of the samples A1, B1, and C1 the coke on the catalyst particle was about the same before and after calcination. The higher apparent coke as a percent of the feed is a result of the higher catalyst to oil required by the calcined samples to attain the given conversion level. In the case of samples D and E with as received reduced vanadium, calcination produces both more activity and much more coke on the catalyst. The implication is that oxidized vanadium on the calcined catalyst is producing additional coke. The results of Table 3 suggest that an oxidized vanadium atom produces about three atoms of carbon.

A similar increase in hydrogen is observed for the samples D and E, Table 4. The results show a similar ratio, about 3 hydrogen atoms are produced per oxidized vanadium. Hydrogen production per gram of catalyst for samples A1,

Table 4. Comparisons of activity and selectivities for hydrogen production for the equilibrium catalysts of Table 2 as-received and after calcination for two hours at 705°C.

Eq. Catalyst	A1	B1	C1	D1	D2	E1	E2
Wt. % Carbon	0.03	0.07	0.08	0.17	0.16	0.31	0.39
ppm V	3325	1557	4424	4172	4266	5733	5859
µmV/gm Catalyst	65	31	87	82	84	113	115
Activity							
% Conversion	60	67	67	60	60	54	54
KC (conv/(1-conv))	1.50	2.03	2.03	1.50	1.50	1.17	1.17
Catalyst/Oil Ratio							
As Received	3.8	3.9	4.7	4.3	3.9	4.5	4
Calcined 2 Hr. 1300F°	4.3	4.6	5	4.1	3.9	3.7	3.5
Hydrogen Selectivity Comparisons, MAT							
As Recieved							
Wt % H2 on Feed	0.25	0.17	0.44	0.18	0.23	0.48	0.44
H2/KC	0.17	0.08	0.22	0.12	0.15	0.41	0.37
µmH2/gm cat	658	436	936	419	590	1067	1100
Calcined 2 Hr. 1300F°C							
Wt % H2 on Feed	0.32	0.23	0.47	0.32	0.32	0.52	0.63
H2/KC	1.82	2.26	2.50	1.82	1.82	1.69	1.80
µmH/gm cat	744	500	940	780	821	1405	1800
Hydrogen Selectivity Differences (Calcined - As Received)							
Wt% H2/KC	1.65	2.18	2.28	1.70	1.67	1.29	1.43
µmH/gm catalyst	86	64	4	362	231	339	700

B1, and C1 does not change with oxidation, consistent with the fact that there is no change in the vanadium oxidation state for these samples.

B. Equilibrium Catalysts Spent (fully coked) and from the Regenerator

Equilibrium catalysts are obtained from the regenerator, and even for the higher coke samples of Table 2 much of the coke has already been burned off and the metals may have been at least partially oxidized. Consequently we have obtained equilibrium catalysts from the stripper, before any coke removal, as well as from the regenerator. A third sample was obtained as before, by calcining a regenerator sample for two hours at 704°C. This experiment allows the comparison of a fully unit reduced sample, a regenerated sample, and a laboratory oxidized sample of the same catalyst from the same unit. The TPR curves, Figure 2, show that the regenerator sample has an intermediate degree of vanadium oxidation. Coke and hydrogen selectivities, Table 5, are worst (most hydrogen and coke) for the calcined catalyst and best for the most reduced stripper sample. The reduced stripper sample also makes more gasoline at a reduced activity.

C. CO Reduction

In both of these experiments the more reduced samples also contain coke. It is possible that the coke already on the catalyst by some selective poisoning mechanism may be responsible for some of the observed selectivity effects. Consequently in a third experiment an equilibrium catalyst, E2 from Table 2, was first oxidized in the laboratory (2 hours, 704°C, air) to

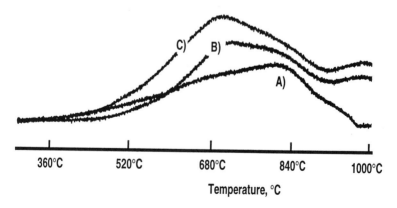

Figure 2. Temperature Programmed Reduction (TPR) of an equilibrium catalyst (1190 ppm V, 906 ppm Ni) including A) coked stripper sample, B) Regenerator sample and C) sample B after air calcination for 2 hours at 705°C.

Table 5. Selectivities of an Equilibrium Catalyst from the Stripper (high coke), the Regenerator (low coke), and after Calcination.

Composition

Al2O3		35.1	
Re2O3		1.48	
Na2O		0.33	
V, ppm		1190	
Ni, ppm		906	
Carbon	0.6	0.05	0
TPR, μm H2/gm Catalyst	90	120	200

Selectivities	Stripper	Regen	Muffle
Conv	53	53	53
C/O	**4.9**	**3.8**	**3.8**
Delta Coke	**0.28**	**0.61**	**0.74**
H2	**0.15**	**0.14**	**0.23**
C1+C2	1.25	1.2	1.2
C3=	3	3	2.9
Total C3	3.5	3.5	3.4
iC4=	1.6	1.3	1.2
C4=	4.9	4.5	4.4
Total C4	6.8	6.9	6.7
Gasoline	**39.5**	**38.5**	**38**
LCO	26.9	25.6	26.2
Bottoms	20.1	21.4	20.8
Coke as % of Feed	**1.5**	**2.3**	**2.8**

remove residual coke and then reduced in CO for two hours at 740°C. The TPR results, Figure 3, show that the reduced and equilibrium catalysts have identical curves, consequently all of the vanadium on the equilibrium catalysts is in the reduced state. The only difference between the equilibrium sample and the CO reduced sample is the presence of 0.3% carbon on the as received sample.

The activity and selectivity results for these samples in Table 6 show that both reduction and coke have an effect on activity. There is an increase in activity with muffle calcination and a further increase with CO reduction. The increase in activity of the CO reduced sample compared to the regenerator sample is due to coke deactivation. The somewhat smaller increase in activity of the reduced sample compared to the calcined sample is apparently due to reduction. The hydrogen and coke selectivities of both reduced samples, regenerator and CO reduced, are similar and much better than the calcined sample. The significant amount of carbon on the regenerator sample compared

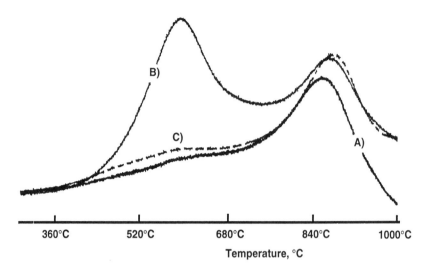

Figure 3. Temperature Programmed Reduction of equilibrium catalyst sample E2 from Table 1 A) as received, B) after air calcination of the as received sample for two hours at 705°C, and C) after CO reduction of the calcined sample for two hours at 705°C.

to the CO reduced sample appears to improve coke selectivity slightly and slightly degrade gasoline selectivity. The gasoline selectivity of the calcined sample was the poorest.

IV. Conclusions

These results show that the oxidation state of vanadium has an important influence on the selectivity of vanadium containing catalysts for hydrogen and coke. It is important to control the oxidation state of a vanadium containing catalyst at the conclusion of a laboratory deactivation, whether cyclic or by impregnation, and before activity and selectivity testing. These results imply that even in the case of riser pilot unit testing, any comparison between the pilot unit and an operating FCCU requires monitoring and control of the oxidation/reduction state of the regenerated catalyst.

These results can also provide guidance in the operation of an FCCU for cracking residual feed stocks containing significant amounts of vanadium. The ideal operation is one that removes as much coke as possible from the spent catalyst and at the same time maintains vanadium in a reduced state.

Table 6. Selectivities of equilibrium regenerator samples as recieved, after calcination, and after calcination and reduction.

Composition			
Al2O3		43.4	
Re2O3		0.93	
Na		0.27	
V, ppm		5859	
Ni, ppm		2947	
Carbon	0	0.39	0

Selectivities	CO Reduced	Regenerator	Muffle
Conv	56	56	56
C/O	3.5	4.4	3.9
Delta Coke	1.21	0.98	1.6
H2	0.45	0.45	0.64
C1+C2	1.55	1.85	1.8
C3=	3.1	3.6	3.2
Total C3	3.6	4.3	3.8
iC4=	1.3	1.6	1.3
C4=	4.6	5.2	4.3
Total C4	6.8	7.4	6.6
Gasoline	39.2	38	36.9
LCO	25.6	25.5	24.8
Bottoms	18.4	18.5	19.2
Coke as % of Feed	4.4	4	6.4

References

1) ASTM Test # D 4463-91, <u>1995 Annual Book of ASTM Standards</u>, Vol. 05.03, p. 862.

2) Nielsen, R. H., and Doolin, P. K. (1993). Fluid Catalytic Cracking: Science and Technology (J. S. Magee and M. M. Mitchell, eds.), <u>Studies in Surf. Sci.</u>, Vol. <u>76</u>, Elsivier, p. 339.

3) Mitchell, B. R. (1980). Metal Contamination of Cracking Catalysists. 1. Synthetic Metals Deposition on Fresh Catalysts, <u>Ind. Eng. Chem. Product Research and Development</u>, **19** 209.

4) Chester, A. W. (1987). Studies on the Metal Poisoning and Metal Resistance of Zeolitic Cracking Catalysts, <u>Ind. Eng. Chem. Res.</u>, **26**, 863.

5) Cimbalo, R. N., Foster, R. L. , and Wachtel, S. J. (1972). Deposited Metals Poison FCC Catalyst, <u>Oil & Gas Journal</u>, May 15, p. 112.

6) Haas, A., Suarez, W., and Young, G. W. (1992). Evaluation of Metals Contaminated FCC Catalysts, <u>Advanced Fluid Catalytic Cracking Technology</u> (K. C. Chuang, G. W. Young, and R. M. Benslay, eds.), AIChe Symposium Series, **291**, Vol. 88, American Institute of Chemical Engineers, New York, p. 133.

7) Benslay, R. M., Chuang, K. C. , Wolterman, G. M., and van de Gender, P. (1992). Evaluating Resid Catalyst with New Matrix Systems Using Pilot Plant Metals Aging Techniques, <u>Advanced Fluid Catalytic Cracking Technology</u>, (K. C. Chuang, G. W. Young, and R. M. Benslay, eds.), AIChe Symposium Series, **291**, Vol. 88, American Institute of Chemical Engineers, New York, p. 120.

8) Boock, L. T. , Petti, T. F., and Rudesill, J. A. (1995). "Recent Advances in Contaminant Metal Deactivation and Metal Dehydrogenation Effects During Cyclic Propylene Steaming of FCC Catalysts," International Symposium on Deactivation and Testing of Hydrocarbon Conversion Catalysts, Division of Petroleum Chemistry Preprints, Vol. 40, #3, p. 421.

9) Gerritsen, L. A., Vervoert, H.N.J., and O'Connor, P. (1991). Cyclic Deactivation: A Novel Technique to Simulate the Deactivation of FCC Catalysts in Commercial Units, <u>Catalysis Today</u>, **11**, 61

10) Bearden, R., and Stuntz, G. F. Passivation of Cracking Catalysts; Nickel, Vanadium or Iron Deposits, U.S. Patent 4,280,896.

11) ASTM Test # D 5154-91, 1995 Annual Book of ASTM Standards, Vol. 05.03, p. 908.

12) Cheng, W. - C., Juskelis, M. V., and Suarez, W. (1993). Reducibility of Metals on Fluid Cracking Catalyst, Appl. Catal. A: General, **103** 87.

13) Woolery, G. L., Chin, A. A., Kirker, G. W., and Huss, A., Jr. (1988). X-ray Absorption Study of Vanadium in Fluid Cracking Catalysts, Fluid Catalytic Cracking, (M. Occelli, ed.), ACS Symposium Series Vol. **375**, American Chemical Society, Washington DC, p. 215.

ACCESSIBLE CATALYSTS FOR SHORT
CONTACT TIME CRACKING

P. O'Connor*, F. P. Olthof*, R. Smeink*,
and J. Coopmans**

*   Akzo Nobel Catalysts, Amsterdam, The Netherlands

**   Akzo Nobel Corporate Research, Arnhem, The Netherlands

SUMMARY

At present there is a strong direction in Fluid Catalytic Cracking (FCC) Technology to reduce the overall contact time between the catalyst and the hydrocarbon molecules.

Short Contact Time (SCT) Cracking seems to be a preferred configuration for new FCC units to be built and for existing FCC units to be revamped aiming at an improved selectivity towards primary products and an improved yield of light (LPG) olefins.

Reaction Temperatures and Catalyst to-Oil (C/O) ratios are sometimes increased in order to compensate for the loss in reaction time, while usually also feed distribution systems are improved and Catalyst Activity levels need to be raised.

As a consequence of this trend towards reduced contact times between catalyst particles and hydrocarbons, the mass transfer rate of hydrocarbons into the catalyst particle ("Ingress") and mass transfer rate out of the catalyst particles ("Egress") will start to play a bigger role even in the case of cracking lighter vacuum gasoils (VGO).

We may therefore expect that catalysts with an improved accessibility (Ingress as well as Egress) which are already effectively applied in Resid FCC will also yield better results in SCT VGO FCC.

I. BACKGROUND AND SCOPE OF WORK

For resid FCC operations mass transfer limitations can be crucial in determining the catalyst performance (O'Connor and van Houtert, 1986) and the accessibility of active sites needs to be optimized (O'Connor et al, 1991).

Based on the same considerations we may expect that also for vacuum gasoil operations, mass transfer limitations may become important when the reaction time is further reduced.

Figure 1 illustrates the trend in FCC catalyst contact times, while an example of the FCC product yield shift after a Short Contact Time (SCT) revamp is given in Table 1. The MSCC (Milli Second Contact time Cracking) Process (Kauff and Bartholic, 1996) is the "extreme" example.

Typically after a SCT revamp gasoline and LPG Olefins yieldsincrease and in some cases bottoms yields tend to deteriorate. Regarding the definition of "Short" in Short Contact Time, Table 2. indicates that the reacting and disengaging time is significantly reduced in SCT revamps/modifications.

In this paper we will discuss the effect of contact time on catalyst performance on a laboratory scale, and demonstrate the benefits of the traditionally more accessible top of the line Resid FCC Catalyst Technology for these operations and the need for testing tools to discriminate between conventional and SCT.

II. EFFECT OF SHORT CONTACT TIME FROM LABORATORY SCALE UNITS

Establishing actual (Catalyst and/or Oil) contact times in laboratory test units as applied in FCC catalyst testing is certainly not an easy task.

Table 1
EXAMPLE
EFFECT OF A SCT REVAMP

	BEFORE	AFTER
Conversion, % vol	77.2	76.8
Dry Gas, %wt	2.9	2.6
LPG,% vol	27.3	28.4
Gasoline, %vol	61.2	60.4
Bottoms, %vol	6.0	8.1
C4 Olefinicity	0.55	0.59
Reactor Temp (ºC)	Base	+ 5
Regenerator Temp (ºC)	Base	-15
Catalyst (ASTM) MAT	Base	+ 4 points

Figure 1

TRENDS IN FCC CONTACT TIMES

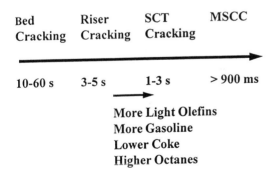

Bed   Riser   SCT   MSCC
Cracking   Cracking   Cracking

10-60 s   3-5 s   1-3 s   > 900 ms

More Light Olefins
More Gasoline
Lower Coke
Higher Octanes

Table 2
WHAT IS SHORT CONTACT TIME?

Example of tracer studies (mean time in seconds)

	Conventional Riser	Example SCT
Feed Injection Zone	1-2	1-2
Reaction Zone (1/3 riser height)	3-5	2-3
Exit Reactor (disengager)	12-15	5-7

Recently Helmsing (1996) constructed an ideal plug flow riser type reactor (Micro Riser, MR) in which test the residence time and time distribution was properly determined. The residence time was varied from 0.7 to 5 seconds. Several types of FCC Catalysts were tested which had been studied in a modified MAT test, the Micro Simulation Test (MST) operating with a total catalyst on oil time (or injection time) of 15 seconds (Table 3).The catalysts were tested in the Micro Riser with a Heavy (Kuwait) Vacuum Gasoil (VGO) feedstock (see Table 4).

Using his data and not correcting for the drop in contact time we can calculate that the second order Cracking rate (Kr, see Table 5), obviously drops significantly for all catalysts.

Table 3
CONTACT TIMES IN TEST UNITS

Test	Estimated mean time (sec)		
	Feed	Catalyst	
MAT	~25	>50	
MST	~ 8	>15	
Pilot Riser	~ 5	~ 5	(PR; Modified ARCO)
Mini Riser	1-5	1-5	(MR; TU Delft)
SCT-1	<5	< 5	(Proprietary test)

Table 4
FEEDSTOCK ANALYSES

Feed	H VGO Heavy VGO	MH Resid Medium Heavy Resid
Density (50 °C)	0.903	0.905
ConCarbon %wt	0.5	4.3
Sulphur %wt	2.93	4.34
Total N ppm	950	1300
HPLC Sat %wt	49.5	53.6
IBP (D1160) °C	370	312
50%vol °C	451	488
70%vol °C	486	549
FBP (HTGC) °C	562	720

Table 5
DEFINITION OF CRACKING RATE

Simplified FCC Kinetics

C = Conversion (%wt) = 100 - LCO - HCO

K_r [s^{-1}] = Cracking Rate = { C/(100-C) } / (CTO. t)

CTO = Catalyst Circulation Rate / Feed Rate

t = an "average" contact time

A drop in the rate constant of at about 50% is observed when going from 5 to 1 seconds in the Micro Riser (Table 6.), this is less than the 80% we would expect from the simplified Cracking rate equation in Table 5, and falls more in line with a lower order of time α in the Voorhies relation:

Coke ~ (function of time) $^\alpha$

Whereby α usually is in the range of 0.25-0.4 as reported in the literature (e.g Helmsing, 1996).

Table 6

EFFECT OF CRACKING TIME

In laboratory test units with H VGO

Test	Average Cracking Time [seconds]	Relative Cracking Rate
MST	~8	100
MR	4-5	80-100*
MR	0.7	30-40*

*) Calculated from data by M.P .Helmsing et al (1996)

This means that also at short contact times there seems to be a rapid deactivation of the catalyst in time resulting in a reduced cracking rate after the first (milli)seconds.

Assuming that the cracking rate is a factor 0.25 order in time, we can calculate that a reduction in contact time from 3 to 1 seconds will result in a reduction in cracking rate of about 25% and hence a loss in conversion of approximately 5% wt at the 75%wt conversion level.

This is reasonable in line with FCC Catalyst targets and modifications for SCT operations, which usually areon the order of 5 points higher conversion and traditionally would mean that at least 25% more active sites are required to reach this conversion.

III. CATALYST EFFECTS AT SHORT CONTACT TIMES

III.1 Catalyst Approach: Active Sites vs Accessibility

Surely increasing the number of active sites in the zeolite and/or active matrix of an FCC catalyst will be advantageous in order to improve the conversion in SCT operations.

We may expect however that as in the case with Resid Operations (O'Connor and Humphries, 1993) at (very) short contact times the accessibility and hence utilization of the active sites can become the key issue (Figure 2).

CATALYST UTILIZATION

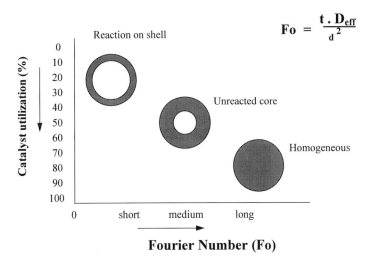

Figure 2

To adress this hypothesis we evaluated several catalysts of about equal activity (active sites) in the MST and in an especially developed Small Scale Short Contact Time laboratory scale test (SCT-1) in which the catalyst and oil contact time is below 5 seconds.

III. 2 Experimental and Results

Case One: Deactivated Catalysts

Three catalysts (conventional , medium resid and top resid) were all deactivated by a Cyclic Deactivation procedure (Gerritsen et al,1991) using 5000 ppm nickel and 1000ppm vanadium.

All three catalysts were tested in a Micro Simulation Test (MST) , being a modified MAT test and in a generic proprietary lab scale Short Contact Time Test (SCT-1). As feeds a medium heavy resid and a (heavy) VGO were applied (See Table 4).

Table 7 summarizes the results in terms of (Relative) Cracking Rates;

Table 7
RELATIVE CRACKING RATES, Kr

Catalyst /Test	MST MH Resid	SCT-1 MH Resid	SCT-1 H VGO
Conventional *	100	100	100
Medium Resid	102	121	104
Top Resid	104	132	110

*) Base set at 100

Table 8
BOTTOMS CRACKING RATES, Kb

BC = Bottoms Conversion (%wt) = 100 - HCO

Kb [1/s] = Bottoms Cracking Rate = { BC/(100-BC)} / (CTO. t)

CTO = Catalyst Circulation Rate / Feed Rate

t = an "average" contact time

    While in the MST even for Resid there is hardly a difference between the conventional and best resid catalyst , at shorter contact times (SCT-1) the best resid catalyst gives a 10 to 30 % higher cracking rate (going from heavy VGO to medium heavy resid).

    If we calculate a bottoms cracking rate analogous to the (conversion) cracking rate (Table 8), the differences become even larger (Table 9), making the top of the best resid catalyst technology a very suitable SCT candidate, as the conversion of bottoms is usually the critical challenge for catalysts in SCT operations.

Table 9

RELATIVE CRACKING RATES
Kr and Kb

Catalyst /Test	SCT-1 MH Resid Kr	SCT-1 MH Resid Kb
Conventional *	100	100
Medium Resid	121	121
Top Resid	132	146

*) Base set at 100

Table 10

MST VS SCT-1 TESTING
Commercial Equilibrium Catalyst Evaluation

Test:	MST				SCT-1			
Feed:	H VGO		MH Resid		H VGO		MH Resid	
Catalyst:	Kr	Kb	Kr	Kb	Kr	Kb	Kr	Kb
A*	100	100	100	100	100	100	100	100
B	124	108	125	110	143	153	155	159

*) Base set at 100%

Case Two: Commercial Equilibrium Catalysts

Two commercial equilibrated FCC catalysts from the same unit: Catalyst A and B were tested in the MST and in the SCT-1. Catalyst B outperformed by far in the commercial unit but this was not obvious from MST testing of the equilibrium catalyst.

Table 10. summarizes the results in terms of relative (conversion) cracking rates and bottoms cracking rates: Indeed at short contact time testing a dramatic difference of about 50% in bottoms cracking rate materializes.

III.3 Consequences for Catalyst Development

From the foregoing results we must conclude that realistic short contact time tests are needed in order to develop and evaluate the appropiate catalysts for SCT operations: Benefits in catalyst accessibility which will be of great importance are not adequately exposed in the existing relatively long injection time laboratory scale fixed-bed tests as MAT (~ 50 seconds) and MST (~15 seconds).

The SCT-1 test is one of the SCT tests now being used in our R&D laboratories to develop the next generation of SCT Catalysts. Tables 11 and 12 demonstrate some preliminary results in improvements in (bottoms) cracking rates which have been observed in this small scale SCT test with experimental catalyst technologies.

IV. ACCESSIBILITY: INGRESS AND EGRESS

From the foregoing we may infer that mass transfer into the catalyst particle is indeed an important factor in the reaction stage of a SCT FCC unit. The same can be posed for mass transfer out of the particle at the end of the reactor and hence in the Catalyst-Hydrocarbon disengaging and/or (pre) stripping stage.

The importance of stripping has been discussed amongst others by O'Connor et al(1991), Gerritsen et al (1991), and Yanik and O'Connor (1995).

Table 11
CATALYST DEVELOPMENTS
SCT-1 Testing, fresh catalyst

Feed:	MH Resid Kr	Slurry Kr
Catalyst System:		
Base	100	100
Improved	108	117
Special Resid	116	225

Table 12
CATALYST DEVELOPMENTS
SCT-1 Testing, fresh vs steamed catalyst

Deactivation:	FRESH	STEAMED*
Feed:	Slurry	Slurry
	Kr	Kr
Catalyst System:		
Base	100	100
Improved	117	132
Special Resid	225	200

*) 5hours 788 °C

CATALYST ARCHITECTURE & ACCESSIBILITY
Ingress and Egress

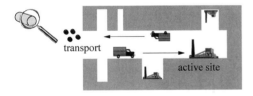

Figure 3

Brevoord et al (1996) demonstrate that catalyst composition and architecture can have a significant impact on adsorbed hydrocarbons as well as on the stripping rate and suggest a correlation with accessibility.

Indeed if our simple model of accessibility as visualized in Figure 3 is correct we would expect some correlation between the Ingress and the Egress, if the Egress is not hampered too much by pore mouth blocking with coke.

If for a broad set of catalyst technologies we correlate "Accessibility" (Ingress) as defined by O'Connor and Humphries (1993) versus "Strippability"

CORRELATION INGRESS VS EGRESS
Accessibility vs Stripping Rate

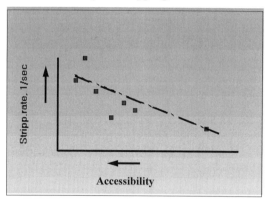

Figure 4

Table 13

ANALOGIES: HARDWARE vs SOFTWARE

GOAL	HARDWARE FCC Unit	SOFTWARE FCC Catalyst
Rapid Vaporization Maximize Catalytic Cracking	Improve Feed dispersion,mixing	Improve Catalyst Accessibility "Ingress"
Reduce Thermal Cracking and Secondary Reactions &	Minimize Post Riser Time	Minimize Cat in/out time
Minimize Hydro carbon carry over to Regenerator	Improve Stripping	Improve Catalyst Strippability "Egress"
	Macro Scale	*Micro(n) Scale*

(Egress) as defined by O'Connor , Pouwels and Wilcox (1992) we find that there is a certain correlation (Figure 4).

It is interesting to note that the foregoing offers some clear analogies between the developments in SCT (Resid) FCC regarding hardware (Unit Design) and software (Catalyst Design), as is demonstrated in Table 13.

At shorter contact times improvements in the process of contacting the hydrocarbons with the catalyst and separating them again will be of greater importance than "traditional" improvements in the reactor/riser process; for Catalyst design this means that Catalyst Architecture and Accessibility will play a greater role.

CONCLUSIONS

Realistic short contact time tests are needed in order to develop and evaluate the appropiate catalysts for SCT operations as benefits in catalyst accessibility which will be of great importance are not adequately exposed in the existing relatively long injection time laboratory scale fixed-bed tests.

For SCT operations higher activity FCC catalysts will be needed.

Mass transfer into (Ingress) and out of (Egress) the catalyst particle is an important factor in the reaction stage of a SCT FCC unit. There is a correlation between the Ingress and the Egress of catalysts as long as the Egress is not hampered by excessive pore blocking with coke.

At shorter contact times improvements in the process of contacting the hydrocarbons with the catalyst and separating them again will be critical. For Catalyst design this means that catalyst architecture and accessibility will play a greater role.

REFERENCES

1. O'Connor,P. and van Houtert,F.W., Paper F-8, Akzo Catalyst Symposium 1986, The Netherlands, H.Th. Rijnten and H.J. Lovink, editors.

2. O'Connor,P.,Gevers,A.W.,Humphries,A.P,Gerritsen,L.A.and Desai,P.H.,"Concepts for Future Residue Catalyst Development", ACS Symposium Series 452, Chapter 20, p. 318, M.L. Occelli, editor, ACS Washington DC 1991.

3. Kauff,D.A.,Bartholic,D.B.,Steves,C.A.and Keim,M.R., "Successful Application of the MSCC process" NPRA Annual Meeting, March 1996 paper AM-96-27.

4. Helmsing,M.P.,Makkee,M. and Moulijn,J.A., "Development of a bench-scale Fluid Catalytic Cracking Microriser" ACS Symposium Series 634, Chapter 24,

p 322. P.O'Connor,T.Takatsuka and G.L.Woolery, editors
ACS Washington DC 1996.

5. Helmsing,M.P., "FCC Catalyst Testing in a novel laboratory riser reactor"
Ph D Thesis, Technical University of Delft, Netherlands 1996.

6.O'Connor,P and Humphries,A.P., "Accessibility of functional sites in FCC"
Preprints ACS vol 38 no.3, 1993; p 598.

7. Gerritsen,L.A. ,Wijngaards, H.N.J.,Verwoert,J. and O'Connor,P.,
Catalysis Today (1991) 11, p. 61.

8.Yanik,S.J and O'Connor,P.,   "Key elements in optimizing catalyst
selections for Resid FCC units" NPRA Annual Meeting, March 1995 paper
AM-95-35.

9. Brevoord,E., Pouwels,A.C.,Olthof,F.P.P.,Wijngaards,H.N.J., and O'Connor,P.,
"Evaluation of Coke selectivity of Fluid Catalytic Cracking Catalysts" ACS
Symposium Series 634, Chapter 25, p 340. P.O'Connor,T.Takatsukaand
G.L.Woolery, editors ACS Washington DC 1996.

10.O'Connor,P., Pouwels,A.C. and Wilcox,J.R., "Evaluation of Resid FCC
Catalysts" AIChE Annual Meeting , November 1992, paper 242E.

EFFECT OF MATRIX ACIDITY ON
RESID CRACKING ACTIVITY OF FCC CATALYSTS

Saeed Alerasool, Patricia K. Doolin, and James F. Hoffman
Ashland Petroleum Company, P. O. Box 391, Ashland, KY 41114

ABSTRACT

The relationship between matrix acidity and resid cracking activity of
FCC catalysts is discussed. This relationship is investigated by correlating both
the virgin and pseudoequilibrium matrix acidity data with resid cracking
activities of various FCC catalysts. Virgin matrix acidity is determined after
complete destruction of the zeolite phase by acid treating the catalyst at a pH of
2. Properties of acid treated catalyst closely approximate those of pure matrix.
However, the virgin matrix acidity does not correlate with resid cracking
performance of the catalyst under commercial or pilot plant conditions. A
second method has been developed that facilitates the measurement of matrix
acidity on a laboratory deactivated sample with properties similar to those in
the commercial cracking unit. The approach is based on total destruction of the
zeolite by steaming at 870°C for five hours. Acidity of the steamed catalyst
closely resembles that of steamed pure matrix and has a good correlation with
the resid cracking activity in the pilot plant testing. Preservation of matrix
acid sites during hydrothermal deactivation is an important requirement for
achieving high resid cracking activity.

The importance of acid type is also discussed. The Brönsted acidity of
the matrix after steaming at 870°C correlates well with resid cracking while
Lewis acid sites correlate poorly. The Brönsted sites are postulated to form by
the interaction between the silica in the binder and active alumina during
hydrothermal deactivation in the commercial FCC unit.

INTRODUCTION

Since the 1970's the average crude oil processed in the cracking units
has been becoming heavier and FCC units have had to crack feedstocks
containing larger resid fractions. According to a NPRA report (1), resid
constituted 15-20% of the total cracker feedstock in 1990. As a result, many
refiners have adapted their crackers to be able to handle such heavy feedstocks
containing higher metal contaminant levels and heavy hydrocarbon molecules.
To accomplish this, not only design and operating changes were made to the
FCC but catalysts have been redesigned such that they can endure the high
metal contaminants of the resid and crack the resid hydrocarbons into valuable
products. Modern cracking catalysts contain REY or USY zeolite embedded in
an inorganic oxide matrix. The zeolite provides most of the cracking activity
while the matrix possesses physical as well as some catalytic functions. (2) The
most important physical role of matrix is its ability to bind the zeolite
crystallites together in a micro spherical catalyst particle hard enough to
withstand interparticle and reactor wall collisions in a commercial catalytic
cracking unit. (3) In addition, matrix with its large number of mesopores
(>20Å) provides a medium for the diffusion of feedstock molecules and cracked
products. It also provides a means for heat transfer during reaction and
regeneration thereby protecting the zeolite structure from structural damage by
increasing its hydrothermal stability. (2, 3, 4) Migration of sodium ions from
zeolite into the matrix at high temperatures has also been suggested as a cause
of improved stability. (3) In addition to the physical functions of the matrix, it
can also play a catalytic role in the cracking reaction. Especially in cracking
applications where resid is mixed with the feed, the large higher boiling point
hydrocarbons in the resid (b.p.>540°C) cannot enter the internal structure of
zeolite. An active matrix with large mesopores can pre-crack these molecules
into smaller fragments before they enter the zeolite and are further cracked into
more valuable products. Pre-cracking of large hydrocarbons (i.e., resids) is
expected to be catalyzed by the surface acid sites of the matrix. Therefore, the
activity of a matrix as a catalytic substrate should be related to the density, type,
and strength of its acid sites. In order to evaluate and predict the resid cracking
performance of matrix, it is important to characterize its acidic properties.
Preferably, this characterization should provide information on the acidity of
the matrix after it has been equilibrated at conditions close to those of the FCC
unit. Although the most convenient way of characterizing matrix acidity is to
measure the acidity of matrix before it is incorporated into the catalyst, this
does not necessarily provide a realistic measure of matrix acidity because
interactions between matrix and zeolite during their hydrothermal deactivation
can lead to modification and even generation of new acid sites. (5)
Furthermore, refiners are usually not provided with the matrix when catalyst

vendors submit their catalyst for evaluation. Therefore, development of a reliable method for an "in situ" measurement of matrix acidity can be a valuable tool both for catalyst manufacturers and refiners.

The object of this study is to investigate the relationship between matrix acidic properties and resid cracking performance of the catalyst under commercial operating conditions. The number and nature of acid sites as well as their hydrothermal stability are investigated in an attempt to develop a bench scale tool capable of accurately predicting commercial resid cracking performance.

EXPERIMENTAL

Catalysts. Twenty catalysts were obtained from five different manufacturers. A number of these catalysts were commercially produced while others were in the exploratory or developmental stage. All twenty catalysts had been well characterized and tested on the pilot scale prior to this study. Selected physical and chemical properties of each catalyst are listed in Table I. Samples of corresponding matrices for five of the cracking catalysts, provided by the vendors, were also examined and included in this study.

Sample Preparation. The virgin matrix acidity was determined on a sample of cracking catalyst which had been treated with 9.0 M sulfuric acid to a final pH of 2.0. The pseudoequilibrium matrix acidity was measured after steaming the catalyst at 870°C. Both of these treatments resulted in complete destruction of the zeolite phase of the catalyst. Detailed procedures are described elsewhere. (6)

Acidity Measurements. A 50 mg aliquot of steamed catalyst or matrix was loaded in the aluminum sample pan of a thermogravimetric analyzer (TGA 951, TA Instruments). The sample was heated to 600°C at a constant heating rate of 10°C/min under a dry nitrogen purge and was held isothermal for two hours. It was then cooled to 100°C and the nitrogen flow was redirected to pass through a saturator containing pyridine (Fisher Scientific, 99.9% purity) before entering the TGA. This configuration allowed the sample to be exposed to a constant flow of nitrogen/pyridine mixture ($P_{pyridine}$ = 20 mm Hg) at an adsorption temperature of 100°C. The adsorption of pyridine continued for approximately 1.5 hours or until no additional weight gain was noted. At the conclusion of this step, the flow of pyridine/nitrogen was discontinued and was replaced with a flow of pure nitrogen at 100°C. This step was required for removing reversibly and physically bound pyridine from the surface. The desorption step continued until no additional weight loss was noted. Values of total acidity were determined from the following equation:

$$\text{Total acidity} = \{[(W_f - W_i)/ W_i]/MW\} \times 10^6 \text{ } \mu\text{mole/g} \quad [eqn. 1]$$

Table I
Selected Properties of Virgin Catalysts

Catalyst	UCS[a]	Z.I.[b]	SA[c]	MSA[d]	%REO[e]	Hg PV[f]
A	2.467	11.2	207	131	0.4	0.42
B	2.465	12.5	252	112	1.5	0.44
C	2.462	13.3	194	109	1.5	0.40
D	2.464	16.1	213	80	2.3	0.43
E	2.462	11.0	210	117	1.7	0.40
F	2.460	12.0	203	69	1.2	0.23
G	2.468	14.3	230	64	1.0	0.28
H	2.470	12.5	157	40	3.3	0.29
I	2.453	7.4	174	128	1.4	0.30
J	24.63	10.9	180	33	1.4	0.31
K	2.467	16.6	159	90	1.4	0.31
L	2.467	17.1	255	118	3.8	0.27
M	2.477	13.6	218	105	3.5	0.36
N	2.456	17.1	230	124	0.1	0.28
O	2.473	13.7	217	124	3.4	0.31
P	2.472	8.5	186	94	3.4	0.40
Q	2.457	8.0	195	111	2.0	0.60
R	2.456	16.3	211	54	2.7	0.27
S	2.461	15.7	203	67	2.6	0.29
T	2.472	8.3	210	114	2.3	0.29

(a): UCS: unit cell size (nm)
(b): Z.I.: Relative zeolite intensity determined by XRD
(c): SA: Total surface area (m²/g)
(d): MSA: Mesopore surface area (m²/g)
(e): %REO: Weight percent rare earth oxide, determined by XRF
(f): Hg PV: Mercury pore volume (cm³/g), determined by mercury porosimetry

where, W_f and W_i correspond to the values of catalyst weight after pyridine desorption and prior to pyridine adsorption, respectively, and MW is the molecular weight of pyridine. The steam treated samples were also characterized by nitrogen BET to determine their total surface area and mesopore surface area. The mesopore area was determined by using a t-plot analysis. (7)

Relative populations of Brönsted and Lewis acid sites were measured by diffuse reflectance infrared spectroscopy (DRIFTS) on both pure matrices and catalysts. The DRIFTS cell (Harrick Scientific Corporation) used for this purpose was equipped with a heater and connected to a gas flow system. An additional thermocouple was installed directly above the sample cup and in direct contact with the sample to accurately control the sample temperature. Other modifications made to both the cell and diffuse reflectance accessory were similar to those reported by Venter and Vannice. (8) Experiments were performed using 20-25 mg sample of the solid acid placed in the DRIFTS cell. The sample was pretreated in a dry nitrogen purge while heating to a final temperature of 400°C and maintained for 2 h before cooling to 40°C and collecting a reference spectrum. After reheating to 100°C under a nitrogen purge, the sample was exposed to pyridine for ten minutes. Pyridine adsorption was accomplished by directing the nitrogen flow through a pyridine saturator and flowing the nitrogen-pyridine mixture over the sample. To remove physically bound pyridine, the sample was purged with nitrogen at 100°C for an additional 2 h period. After cooling the sample to 40°C, a spectrum was collected. The reference spectrum recorded from the calcined sample was subtracted from the spectrum collected following the desorption of pyridine at 100°C. Baseline corrections were performed on spectra and relative populations of Brönsted and Lewis acid sites were determined from the absorbance of bands observed at 1542-1545 and 1448-1455 cm^{-1}, respectively.

Pilot Plant Evaluation. To evaluate the methods discussed in this study, the matrix acidity data were correlated with the existing pilot plant performance results. The pilot scale data were gathered over a nine-year period. The operating conditions of the pilot plant unit were selected carefully to mimic those of the commercial reduced crude conversion (RCC®) unit. The RCC® is a patented, fluidized catalytic cracking unit specifically designed to process heavy feedstocks. Among the performance curves constructed from the pilot plant test, the plot of weight percent light cycle oil yield to weight percent slurry oil yield ratio (LCO/SO) as a function of volume percent conversion was chosen as the best indicator of resid cracking capability. Although this ratio is calculated for a range of conversions during the pilot plant evaluation, a single value of LCO/SO at 75% conversion is used in this study because the conversion achieved in the catalytic cracking unit is in the 70-80% range. A detailed description of the pilot plant setup and test conditions is found elsewhere. (9, 10)

RESULTS AND DISCUSSION

The method for measuring the acidity of virgin matrix makes use of the fact that the crystal structure of the zeolitic portion of a cracking catalyst can be completely destroyed by treatment with a concentrated acid without significantly modifying the acidic and physical properties of the matrix. The measured acidity on an acid treated cracking catalyst should, therefore, correspond to the acidity of its pure matrix phase. To be able to rely on the proposed method for the determination of matrix acidity, it should be demonstrated that the values obtained closely approximate the acidity of the matrix component of the virgin catalyst. This has been demonstrated elsewhere. (6)

Once the virgin matrix acidity of all catalysts were determined, the relationship between these values and resid cracking in pilot plant testing was examined. When the values of virgin matrix acidity were plotted against the resid cracking performance of the catalysts, represented by the LCO/SO ratio, virtually no correlation was observed between matrix acidity and resid cracking under pilot plant testing conditions as shown in Figure 1. These findings imply that virgin matrix acidity cannot predict the resid cracking of the catalyst.

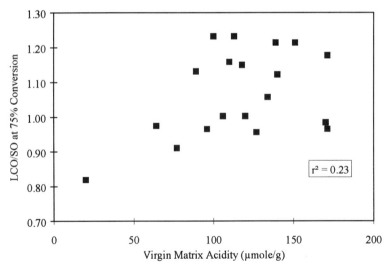

Figure 1: Relationship between virgin matrix acidity and resid cracking

We then postulated that, due to differences in the hydrothermal stability of different matrices, acidity measurement should be performed on a catalyst sample that has experienced some of the severe conditions existing in the commercial unit. Since under actual RCC® conditions, the catalyst is exposed to steam, nickel, vanadium, iron, sodium, and heat, it is logical to assume that its physical and acidic properties will change as a result of being exposed to such adverse conditions. In addition, the catalysts used in this study were from five different manufacturers. These catalyst manufacturers use a variety of preparation methods and compositions including both incorporated catalysts and those prepared by in-situ methods. As a result, in all likelihood, different catalysts will have various degrees of stability to hydrothermal treatment and metal deactivation. To account for the hydrothermal deactivation of matrix, the matrix acidity was determined on catalyst samples steamed at 870°C. This severe steaming leads to total destruction of the zeolite and leaves behind a matrix with physical and acidic properties very similar to that of the commercial equilibrium catalyst. Surface area of matrices from catalysts steamed in the laboratory to simulate an equilibrium matrix were found to be similar to those of RCC equilibrium catalysts obtained from the commercial RCC unit. Since surface area/acidity relationships are linear for each catalyst system, the acidity of the matrices are also similar. Acidity of the laboratory steamed matrix is referred to as "pseudoequilibrium matrix acidity". Figure 2 shows the linear relationship ($r^2 = 0.85$) between resid cracking observed in the pilot plant testing of catalysts, represented by the LCO/SO ratio at 75% conversion, and the pseudoequilibrium matrix acidity. This is in marked contrast with the lack of dependence observed between virgin matrix acidity and resid cracking (Figure 1). These data confirm that an active matrix not only needs to be acidic, it should also be able to preserve a significant fraction of its acidity even after being exposed to such severe conditions as those encountered in the commercial scale process. Since the pseudo-equilibrium matrix acidity accounts for both stability and acidity, it provides a reasonably accurate prediction of resid cracking performance. The pseudoequilibrium matrix acidity method can be a valuable tool for quick and inexpensive evaluation and screening of catalysts to predict their resid cracking activity before performing more time-consuming and expensive pilot plant tests.

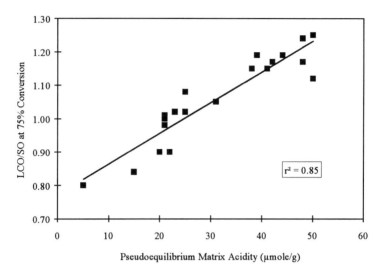

Figure 2: Relationship between pseudoequilibrium matrix acidity and resid
          cracking.

To further understand the role of matrix acidity in resid cracking, the nature of
acid type was also investigated. This was accomplished by performing diffuse
reflectance infrared spectroscopy (DRIFTS) on samples of virgin matrix as well
as catalysts steamed at 870°C after exposing each catalyst sample to pyridine.
Table II contains the relative population of Brönsted and Lewis acid sites on
five catalyst samples and on their corresponding pure matrices. These are the
only catalysts used in this study for which the corresponding pure matrix were
available.
        As shown in Table II, the matrix of catalyst A in its fresh form contains
only Lewis acid sites. Catalyst A has good resid cracking activity and its
superior performance in this respect is believed to be due to its active matrix.
At first glance it appears that Lewis acidity of the matrix of this catalyst is
responsible for its superior bottoms upgrading capability. However, the
"pseudoequilibrium matrix" prepared by steaming at 870°C shows the presence
of Brönsted acid sites. In fact, 36% of all acid sites that remain on the matrix
after severe hydrothermal deactivation are of Brönsted type. A somewhat
similar distribution of acid sites was observed on the severely steamed sample
of Catalyst A. Since steaming at 870°C leads to total destruction of the zeolite
phase, it is not surprising that the steamed catalyst and its corresponding
steamed matrix have similar distributions of Brönsted and Lewis sites. Similar

changes were observed on the other catalysts and their matrices after steaming. (Table II)

Based on these findings, it was concluded that steaming leads to formation of new Brönsted acid sites on the matrix. The exact origin of such sites and the mechanism of their formation are not well understood. However, it is likely that Brönsted sites are formed as a result of the interaction between the silica phase present in the binder and active alumina used in the formulation of the matrix.

Figure 3 shows the relationship between the absorbance of the Lewis band (normally found between 1445 to 1455 cm^{-1}) and the LCO/SO ratio obtained from the pilot scale testing of all twenty catalysts used in this study. A significant amount of scatter is found in these data. The weak correlation (r^2 = 0.46) indicates that resid cracking is not directly affected by Lewis acidity. It is believed, however, that this type of acidity does affect other aspects of catalyst performance. In particular, coke formation is expected to occur through Lewis acid catalyzed reactions; i.e., Lewis acid sites may contribute to non-selective cracking.

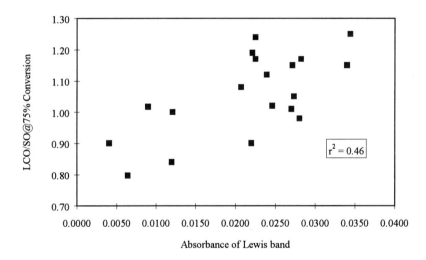

Figure 3: Dependence of resid cracking on Lewis acidity. Data obtained from absorbance of the band at 1450 cm^{-1} after desorption of pyridine at 150°C.

Table II
Effect of Steaming on Matrix Acid Type

| | %Lewis (%Brönsted) | |
Catalyst	Virgin	Steamed @ 870°C
Matrix A	100(0)	64(36)
Catalyst A	--	72(28)
Matrix P	72(28)	53(47)
Catalyst P	--	61(39)
Matrix N	76(24)	63(37)
Catalyst N	--	54(46)
Matrix U	74(26)	63(37)
Catalyst U	--	47(53)
Matrix H	52(48)	0(0)
Catalyst H	--	0(0)

Figure 4 shows the relationship between the absorbance of the Brönsted band (typically found between 1540 to 1545 cm^{-1}) and LCO/SO ratio. In contrast with Lewis acidity, Brönsted acidity seems to have a much stronger correlation with bottoms cracking ($r^2 = 0.76$). Although the correlation between Brönsted acidity and LCO/SO is not stronger than that between LCO/SO and total pseudoequilibrium matrix acidity, these data do reflect a relationship between Brönsted acidity and resid cracking which is much more pronounced than that between Lewis acidity and resid cracking. This suggests that not only a significant number of acid sites need to survive hydrothermal deactivation, but in all likelihood, another requirement exists for a good resid cracking matrix. Namely, the matrix should be prepared in such a way that a large number of Brönsted sites are created as the catalyst undergoes hydrothermal deactivation in the unit. Although it is not fully understood how this phenomenon can be controlled during the preparation of the catalyst, one possible way may be to use smaller particles of active alumina in order to maximize their interaction with the silica binder. Catalyst vendors might be able to find other ways of maximizing the creation of Brönsted acidity during hydrothermal deactivation.

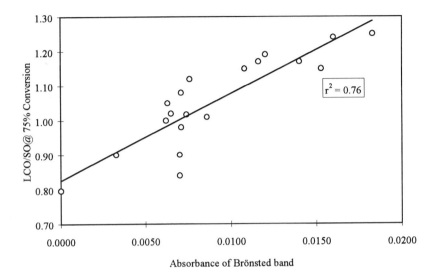

Figure 4: Dependence of resid cracking on Brönsted acidity. Data obtained from absorbance of the band at 1545 cm^{-1} after desorption of pyridine at 150°C.

CONCLUSIONS

Acid treatment and steaming both lead to total destruction of the zeolitic component of the cracking catalyst. Thermogravimetric measurement of acidity by pyridine adsorption on an acid treated cracking catalyst provides an accurate estimate for virgin matrix acidity. Although this was shown to be a reliable method for determining virgin matrix acidity, only a weak correlation was found between the measured values and the resid cracking activity of the catalyst. Similar pyridine adsorption measurements on catalysts steamed at 870°C provided good estimates for the equilibrium matrix acidity of the catalysts. These values were found to have a linear relationship with resid cracking activity of the catalyst. Therefore, the pseudoequilibrium matrix acidity method can be used to estimate the resid cracking of a catalyst. Since the method is quick and inexpensive, it can be a valuable screening tool for both refiners and catalyst manufacturers.

The nature of matrix acid sites was also shown to affect the resid cracking performance of the catalyst. While the total acidity of fresh matrix does not correlate with resid cracking activity of the FCC catalyst, the pseudoequilibrium matrix acidity correlates well with resid cracking under pilot plant and commercial conditions. This is because both hydrothermal stability and high initial acidity are required characteristics of a robust matrix. Catalysts which generate large populations of Brönsted acidity on their matrices during the deactivation process appear to have higher resid cracking activity. The Brönsted acids were postulated to form during hydrothermal deactivation through the interaction between the silica binder and active alumina phase of the matrix.

REFERENCES

1.   NPRA Survey of US Gasoline Quality and US Refining Industry Capacity to Produce Reformulated Gasolines, Part A (1991).

2.   Scherzer, J. Octane-Enhancing Zeolite FCC Catalysts: Scientific and Technical Aspects. *Catal. Rev. Sci. Eng.* **1989**, *31*, 215.

3.   Plank, C. J. The Invention of Zeolite Cracking Catalysts - A Personal Viewpoint. *ACS Symp. Series.* **1983**. *222*, 253, .

4.   Rajagopalan, K. and Habib, E. T. Understand FCC Matrix Technology. *Hyd. Proc.* **1992**, *71*(9), 43.

5.   Suzuki, I., Oki, S., and Namba, S. Determination of External Surface Area of Zeolites. *J. Catal.* **1986**, *100*, 219.

6.   Alerasool, S., Doolin, P. K., and Hoffman, J. F. Matrix Acidity Determination: A Bench Scale Method for Predicting Resid Cracking of FCC Catalysts. *Ind. Eng. Chem. Res.*, **1995**, 34(2), 434.

7.   Anderson, J. R. and Pratt, K; C. *Introduction to Characterization and Testing of Catalysts*, Academic Press, Australia, 1985.

8.   Venter, J. J. and Vannice, M.A. Modifications of a Diffuse Reflectance Cell to Allow the Characterization of Carbon-Supported Metals by DRIFTS. *Appl. Spectry.* **1988**, *42*, 1096.

9.   Mitchell, M. M., Jr. and Moore, H. F. Protocol Development for Evaluation of Commercial Catalytic Cracking Catalysts. *ACS Div. Pet. Chem Preprints* . **1988**, 33(4), 547.

10.   Mitchell, M. M., Jr., Hoffman, J. F., and Moore, H. F. Residual Feed Cracking Catalysts. in *Stud. Surf. Sci. Catal.* **1993**, *76*, (J. S. Magee and M. M. Mitchell, Jr., ed.) 293.

HIGH METAL TOLERANCE MATRIX FOR FCC CATALYST

Jin-Shan Wang, Bing-Lan Li, Xing-Zhong Xu, and Xing-Min Ke

Research Institute of QiLu Petrochemical Co.
Zibo, Shandong, 255400, People's Republic of China

ABSTRACT

The study on the passivation of nickel and vanadium deposited on the FCC catalyst by modified alumina and CBO_3, a synthesized complex oxide, shows that the matrix of the FCC catalyst containing modified alumina and CBO_3 possesses high metal tolerance ability. The activity of the FCC catalyst prepared by using the matrix containing modified alumina and CBO_3 can be increased by 4.3–4.7 units and the specific coke formation can be reduced by 35.5–40.7% compared to that of the catalyst without modified alumina and CBO_3.

INTRODUCTION

In recent years, the FCC unit is always used for processing residue oil containing metals or VGO blended with a certain amount of residue. The heavy metals deposited on the surface of the catalyst during the FCC operation affect the properties of the catalyst. The main heavy metals that affect the properties of FCC catalyst are nickel and vanadium. In general, the structure of zeolite is not affected by nickel. But nickel can catalyze the dehydrogenation reaction of hydrocarbons, promote coke formation on the catalyst surface, and then cause the deactivation of the reaction by covering the active sites or blocking the pores of the catalyst. The activity of the reactivated catalyst may be recovered by air or airstream regeneration; therefore, the nickel poisoning is reversible. As for vanadium, it cannot only promote the coke formation reaction but can also destroy the structure of the zeolite by dealumination of the zeolite framework; therefore the poisoning of vanadium is irreversible. Shien

Jen Yang et al. (1) reported that there is an interaction between nickel and vanadium, which leads to the inhibition effect for vanadium and nickel on the catalyst. Nickel can inhibit the destruction of the structure of USY zeolite caused by vanadium and vanadium can suppress coke formation due to the presence of nickel to some extent (1). Even so, the development of an FCC catalyst having metal passivation function still plays an important role especially for the FCC catalyst used for processing of feed containing a high level metal deposition. In order to passivate the effect of metals on the catalyst, it is very important to develop a high nickel and vanadium tolerance matrix of FCC catalyst because most of the heavy metals are deposited on the surface of matrix of the catalyst during the FCC operation. If the effect of metals on the catalyst can be inhibited by the metal tolerance matrix, the activity and the coke formation selectivity of the catalyst may be improved.

The study of the passivation of nickel and vanadium by phosphor modified alumina and CBO_3, a synthesized complex metal oxide, shows that the activity of the semicommercial FCC catalyst prepared by using the matrix containing modified alumina and CBO_3 can be increased by 4.3–4.7 units and the specific coke can be decreased by 35.6–40.7% compared to that of the catalyst without phosphor modified alumina and CBO_3.

EXPERIMENTAL AND RESULTS

The Modification of Alumina as Nickel Passivation Component

The coke deposited on the surface of the catalyst is partly caused by the presence of nickel that can catalyze the dehydrogenation and coke formation reactions. It is known that active alumina can react with nickel to form nickel aluminate ($NiAl_2O_4$, Spinel), which does not affect the performance of the catalyst. Therefore adding a certain amount of alumina into the matrix of the FCC catalyst may reduce the dehydrogenation and coke formation activities of nickel deposited on the FCC catalyst. But the acid sites of active alumina are mainly the Lewis acid with different strength, which can promote coke formation on the catalyst (2). For this reason, it is important to modify the property and the strength of the acid sites of alumina and thereby to increase its activity and coke formation selectivity when using the modified alumina as a nickel passivation component of the matrix of FCC catalyst.

Phosphoric acid can react with the Lewis acid sites of alumina according to the following reactions:

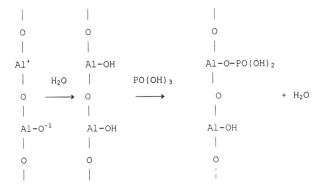

The Lewis acid site first reacts with H_2O to form a hydroxyl group, then reacts with phosphoric acid, and changes into a weak acidic hydroxyl group of the phosphoric acid group. Therefore, the modification of alumina with phosphoric acid cannot only change the Lewis acid sites into Bronsted acid but can also accommodate the distribution of the acid strength. Phosphor modified alumina is also active for the preliminary cracking of large molecular hydrocarbons. Therefore the activity of the catalyst may also be increased by adding phosphor modified alumina to the matrix of the FCC catalyst.

A certain amount of γ-alumina was treated with phosphoric acid solution, filtered, and calcinated at 650°C. Phosphor modified alumina (P-Al_2O_3) samples with different phosphor content were obtained. The quantities of acid sites with different strength were determined by the TPD of ammonia. The quantities of weak acid, medium acid, and strong acid are represented by the amount of NH_3 desorbed at the temperature of 140–260°C, 260–350°C, and 350–650°C respectively. The relative quantities of the Lewis acid and Bronsted acid were determined by the infrared spectrometric method of pyridine absorption at the wave number of 1450 cm^{-1} and 1540 cm^{-1} respectively. The results of the experiments are shown in Figure 1 and Table 1.

From Figure 1 we can see that the quantity of strong acid decreases steadily with the increase of phosphor content. The quantities of weak acid and total acid decrease firstly and then increase with the increase of phosphor content. But the medium acid increases with the increase of phosphor content firstly and then decreases. This phenomenon shows that the acid strength distribution of γ-Al_2O_3 can be regulated by phosphor modification to reduce the quantity of strong acid and to increase the quantity of medium acid. This is advantageous when using phosphor modified alumina as a nickel passivation component of the matrix of the FCC catalyst, because the lowering of strong acid sites can lead to the reducing of coke formation.

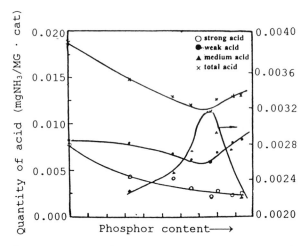

Fig.1 The relationship of acid strength with
the content of phosphor in P-Al₂O₃

In order to evaluate the function of P-Al$_2$O$_3$ on the property of FCC catalyst, a series of model catalysts were prepared by mixing 35% USY zeolite (silica/alumina molar ratio = 10.3) with the matrix comprising kaoline clay and P-Al$_2$O$_3$ with different phosphor content, then dried and calcined at 650°C.

The cracking activities of the model catalysts were characterized by n-hexane cracking rate on the catalyst at 450°C, H$_2$/nC$_6$ mole ratio = 2.9, V =

Table 1. The Relationship Between the Property of Acid and the Property of the Catalysts with Different Phosphor Content

Sample No. (Increase in phosphor content	Relative quantity of B-Acid (A.S/m)	Relative quantity of L-Acid (A.S/m)	$\dfrac{\text{L-Acid}}{\text{(B-Acid)}}$	$\gamma_{c,av}$ (mg/min.gcat)
0 (No phosphor)	1.7	5.7	3.4	1.72
1	1.7	4.2	2.5	1.65
2	1.7	3.8	2.3	1.36
3	1.6	3.7	2.3	0.92
4	2.1	4.0	1.9	0.91
5	2.0	3.6	1.8	1.18
6	1.6	3.5	2.2	1.45
7	1.4	3.7	2.6	2.10

A—IR absorption rate, m—weight of the sample, mg S—Surface of the sample, cm^2
$\gamma_{c,av}$—average coke formation rate.

4h^{-1} in a microreactor. The coke formation selectivity was determined by the average coke formation rate of n-heptane using a thermal balance at 420°C, H$_2$/C$_7$ molar ratio = 4.4. The results are shown in Figure 2 and Figure 3.

The data of Figure 2 and Figure 3 show that the activities of the model catalysts increase and the coke formation rates decrease with the increase of phosphor content. But the activities begin to decrease and the coke formation rates begin to increase when the two curves reach their peak.

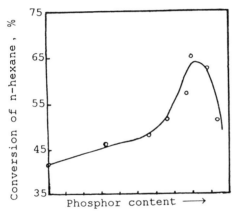

Fig.2 The relationship between the initial activity and phosphor content of the model catalysts

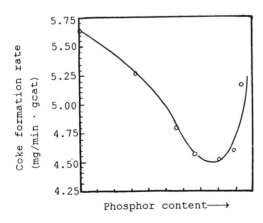

Fig.3 The relationship between the coke formation rate and the phosphor content of the model catalysts

Except that the acid strength of alumina can be accommodated by phosphor treatment, the improvement of the activity and coke selectivity of the catalyst may also be contributed to be reducing the Lewis acid sites (the coke formation centers) when using P-Al$_2$O$_3$ as a component of the matrix. This relationship is shown in Table 1.

It can be seen from Table 1 that as the phosphor content increases, the relative quantity of Lewis acid sites and the ratio of L-Acid to B-Acid is reduced and then maintained about a certain value, the $\gamma_{c,av.}$, decreasing firstly and then increasing. These results demonstrate the interaction between the Lewis acid sites and phosphoric acid as discussed above. The reducing of Lewis acid helps reduce coke formation.

The Study of the Passivation of Vanadium by CBO$_3$

The feed of an FCC unit, especially from the crude of Middle East, always contains a large amount of vanadium, which can deposit on the surface of the catalyst during the FCC operation. Vanadium on the catalyst can be oxidized to V^{+5} (V$_2$O$_5$) during regeneration, which can react with H2O vapor to form volatile vanadic acid (H$_3$VO$_4$):

$$V_2O_5 + 3H_2O \rightarrow 2H_3VO_4$$

The structure of zeolite can be destroyed by the reaction of H$_3$VO$_4$ with framework alumina or by the hydrolysis of zeolite catalyzed by H$_3$VO$_4$, which causes the decrease of crystallinity of zeolite and then leads to the decrease of catalyst activity (4,5). In order to passivate the destructive action of vanadium on zeolite, it is necessary to choose a substance which can easily react with vanadic acid and has no undesired catalytic function. Alkaline earth oxide can react with vanadic acid. But alkaline earth oxide compounds are basic and have a negative effect on the catalytic cracking reaction. The best way to prepare a compound is by the reaction of alkaline earth oxide with a solid acid without undesired effect; the acidity of it should be weaker than that of vanadic acid. CBO$_3$ prepared from the reaction of CO and BO$_2$ may be preferred as a matrix component for the passivation of vanadium. CBO$_3$ can react with V$_2$O$_5$ to form CV$_2$O$_6$:

$$CBO_3 + V_2O_{5m} \rightarrow CV_2O_6 + BO_2$$

Here, C is an alkaline earth metal, B is a metal element, and the acidity of its oxide BO$_2$ is weaker than that of V$_2$O$_5$. In this case, V$_2$O$_5$ on the catalyst is able to react with CBO$_3$, displace the BO$_2$ from CBO$_3$, and forms a new oxide (CV$_2$O$_6$) with low or nonactivity to damage the structure of zeolite.

FCC Catalyst Prepared with the New Metal Tolerance Matrix

According to the above studies, P-Al$_2$O$_3$ and CBO$_3$ are the preferable components of the new metal tolerance matrix of FCC catalyst for the passivation of nickel and vanadium deposited on the catalyst during FCC operation. In order to evaluate the passivation effect, a series of catalysts were prepared.

The catalyst B prepared in a bench apparatus contains USY Zeolite, P-Al2O3, CBO$_3$, kaoline clay, and adhesive agent. For comparison, the catalyst B$_2$ containing the same amount of USY zeolite but no P-Al$_2$O$_3$ and CBO$_3$ was also prepared in the same bench apparatus. The prepared catalyst samples were polluted by adding a certain quantity of nickel nitrate or ammonium vanadate solution containing a proper amount of nickel or vanadium which are equivalent to 3000 ppm Ni or 5000 ppm V on the catalyst samples, then dried and calcinated at 550°C. The performances of the two catalysts loaded with 3000 ppm nickel and 5000 ppm vanadium separately are listed in Table 2 and Table 3. A commercial catalyst C having the same components of catalyst-B$_2$ is also listed in Table 2 and Table 3. The samples used for microactivity test (MAT) were aged by hydrothermal treatment with 100% steam at 800°C for 4 hours. Comparing the data of nickel loaded catalysts in Table 2, we can see that the MA of the Cat-BNi is 4.5 and 8.3 units (5.2% and 10.0%) higher

Table 2. The Performance of Nickel (3000 ppm) Loaded FCC Catalysts

Catalyst	Form	MA (%)	Coke (mg/g.cat)	Specific coke, coke/MA (× 100)
BNi	Bench	84.9	6.09	7.18
B$_2$Ni	Bench	80.4	6.01	7.47
CNi	Commercial	76.6	6.37	8.32

MA—Microactivity.

Table 3. Performance of Vanadium (5000 ppm) Loaded FCC Catalysts

Catalyst	Form	MA (%)	Coke (mg/g.cat)	Specific coke (coke/MA × 100)
BV	Bench	77.7	3.74	4.47
B$_2$V	Bench	76.0	3.83	5.04
CV	Commercial	61.3	3.44	5.62

Table 4. Performance of the Catalysts Produced in a Semicommercial Plant

Catalyst	Ni content (ppm)	V content (ppm)	MA (%)	Coke (mg/g.cat)	Specific coke (coke/MA × 100)
S	—	—	73.7	5.54	7.52
S$_2$	—	—	69.0	8.75	12.68
SNiV	3000	5000	58.7	5.3	9.03
S$_2$NiV	3000	5000	54.4	7.65	14.06

than that of the cat B$_2$ Ni and cat CNi. The specific coke of Cat-BNi is 0.29 and 1.14 units (3.8% and 13.7%) lower than that of the Cat-B$_2$H Ni and Cat-CNi.

The MA of Cat-BV is 1.7 and 16.4 units (2.2% and 26.7%) higher than that of Cat-B$_2$V and Cat-CV. The specific coke of cat-BV is 0.57 and 1.15 units (11.3% and 20.5%) lower than that of Cat-B$_2$V and Cat-CV.

Catalyst Cat-S and Cat-S$_2$ were produced in a semicommercial catalyst production plant. Cat-S contains P-Al$_2$O$_3$ and CBO$_3$ as metal tolerance components; Cat-S$_2$ contains no metal tolerance components. The performances of them are shown in Table 4. Cat-SNiV and Cat-S$_2$NiV are the Cat-S and Cat-S$_2$ polluted with 3000 ppm Ni and 5000 ppm V by the method described above.

The data in Table 4 show that the microactivities of the Cat-S and Cat-SNiV are 4.7 and 4.3 units (6.8% and 7.9%) higher than that of the Cat-S$_2$ and Cat-S$_2$NiV, and the values of specific coke of the Cat-S and Cat-SNiV are 5.16 and 5.01 units (40.7 and 35.6%) lower than that of Cat-S$_2$ and Cat-S$_2$NiV respectively. The data in Tables 2, 3, and 4 demonstrate that the performances of the catalysts containing metal tolerance components P-Al$_2$O$_3$ and CBO$_3$ are better than the catalysts comprising no P-Al$_2$O$_3$ and CBO$_3$. That means, P-Al$_2$O$_3$ and CBO$_3$ are the preferable nickel and vanadium passivation components for the production of FCC catalyst used for the processing of metals containing feed.

CONCLUSION

1. Through a series of studies, a new matrix of FCC catalyst containing P-Al$_2$O$_3$ and CBO$_3$ has been developed that possesses high metal tolerance ability.

2. The Lewis acid sites (coke formation centers) and the acid strength of active alumina can be regulated by phosphor treatment. The phosphor

modified alumina may be used as the nickel passivation component of FCC catalyst. It is also active for preliminary cracking of the feed which can improve the activity of the catalyst.

3. CBO_3, a complex oxide of alkaline earth and metal, possesses vanadium passivation function. It can fix V_2O_5 into CV_2O_6 which is not active to attack the framework of zeolite.

4. The FCC catalyst containing the new metal tolerance matrix $P-Al_2O_3$ and CBO_3 possesses high activity and coke formation selectivity. The performance of the semicommercially produced FCC catalyst comprising the new matrix shows that the microactivities are 4.7 and 4.3 units higher and the specific cokes are 40.7% and 35.6% lower than that of the catalysts without metal tolerance matrix. That means the new matrix has high metal passivation ability. It can be used to produce FCC catalyst for the processing of heavy metals containing feeds such as residue oil or VGO blended with residue oil.

REFERENCES

1. Shien-Jen Yang, Yu-Wen Chen, *Zeolite 15*, 77 (1995).
2. Laurance B, Bright NPRA Annual meeting, Am-91 53 (1995).
3. Stanilaus A Absi, *Appl Catal, 39*, 239 (1988).
4. Occelli M.L., *Catal. Rev. Sci. Eng. 33*, 241 (1991).
5. Wormsbecher R.F., *Catal. 100*, 130 (1986).

INFLUENCE OF THE NATURE OF FCC FEED ON THE PRODUCTION OF LIGHT OLEFINS BY CATALYTIC CRACKING

Th. Chapus, H. Cauffriez, and Ch. Marcilly

Institut Français du Pétrole, 1 & 4 avenue de Bois-Préau, BP 311, 92506 Rueil-Malmaison Cedex, France

1 INTRODUCTION

The 1990 Clean Air Act has set rules for gasoline reformulation that requires major compositional changes, including a higher contribution of oxygenated compounds to the gasoline pool. This explains why FCC units are expected to play a major role in the coming years as a producer of light olefins (propylene, butenes, and amylenes) to be used as feedstock for oxygenate (MTBE/TAME) production.

If it is well known that the nature of FCC feed has a very big impact on the production of gasoline by catalytic cracking, the number of data published in the literature concerning the influence of the nature of the feed on the production of olefins is rather limited (1–6). A priori, based on the chemical composition, aromatic feedstocks have lower olefin potential than naphthenic and paraffinic feedstocks. They have not been studied in this work, which aims at comparing abilities of naphthenic and paraffinic feedstocks to yield light olefins.

2 EXPERIMENTAL

The impact of the nature of FCC feedstock on light olefins production (C3–C5 olefins) has been studied using a MAT unit comprising a fixed-bed reactor containing 5 g of FCC catalyst. WHSV is kept constant, approximately 40 h^{-1}. C/O ratio is varied in the range 3–9, by modifying the time on stream (10–50 s), keeping constant the feed flowrate and contact time as well.

Three real paraffinic and naphthenic feedstocks have been evaluated at 530°C in a MAT unit: a paraffinic feed composed of a mixture of paraffinic resid and VGO, a paraffinic resid, and a naphthenic vacuum gas-oil (VGO). Their main characteristics are given in Table 5. The naphthenic VGO and the paraffinic VGO + Resid have low ConC and high hydrogen content, contrarily to the paraffinic resid. These feeds were evaluated using an equilibrated FCC catalyst (catalyst A) from an industrial Resid Catalytic Cracking Unit (R2R). This catalyst has no significant matrix activity and contains around 5,000 wt ppm metals.

In order to evaluate the maximum olefins production that can be reached by catalytic cracking of a paraffinic feedstock, a pure paraffinic feed (hard wax 48–52°C) was also tested, varying the temperature in the range 530–665°C, and using catalyst B. This catalyst is a marketed FCC catalyst, chosen among low rare-earth containing catalysts in order to limit hydrogen-transfer reactions responsible for consumption of olefins produced by cracking. In order to bring the catalyst activity to a representative level, the fresh catalyst was steamed during 15 hours at 730°C prior to testing. Main characteristics of both catalysts A and B used in this study are shown in Table 6.

3   RESULTS AND DISCUSSION

3.1   Comparative Evaluation of the Naphthenic VGO and the Paraffinic VGO + Resid Feedstocks

Crackabilities and selectivities of these feedstocks were compared at the same conversion (70 wt %) and at the same coke yield (6 wt %). As shown in Table 1, the paraffinic VGO + Resid is much easier to crack than the naphthenic VGO, since the same conversion level (70 wt %) is reached with a much lower C/O ratio (C/O = 3.8 versus 7.1 for the naphthenic feed at 70 wt % conversion). In addition, the paraffinic feed gives, at same conversion, a significantly lower coke yield (3.6 versus 5.4 wt % at 70 wt % conversion). Yield structure at 70 wt % conversion shows that the paraffinic feed produces less dry gas (–35% rel.), more gasoline (+2.5 points), and less LCO (–2.9). Bottoms upgrading is lower for paraffinic feed (–2.6 points), due to the lower C/O ratio. LPG yield is the same in both cases. No difference is observed on propylene, normal butenes, and isobutane yields. Only isobutylene yield is higher with the paraffinic feed (+17% rel.).

C3 olefinicity is not significantly different at the same conversion, and C4 olefinicity is slightly higher for paraffinic feed (0.66 versus 0.64 at 70 wt % conversion). Data given in Table 3 show that paraffinic feed gives at 75 wt % conversion both a slightly higher C5 olefinicity (0.63 versus 0.62), and higher pentenes yield (+20% rel.).

Comparison at the same conversion (data respectively given at 70 and 75 wt % conversion in Tables 1 and 3) show that the paraffinic feed allows

Table 1. Performances and Olefins Yields on Paraffinic and Naphthenic Feedstocks (MAT Results - T = 530°C)

	Comparison at same conversion level (~70 wt %)		
Nature of the Feed	VGO	VGO + Resid	Resid
Character	Naphthenic	Paraffinic	Paraffinic
C/O	7.1	3.8	2.3
Conversion	69.9	70.2	69.9
Coke (wt %)	5.4	3.6	4.4
Delta-coke	0.76	0.95	1.9
H2 + C1-C2	2.3	1.5	2.0
LPG	14.9	15.3	12.5
Gasoline (C5-221°C)	47.3	49.8	51.0
LCO (221-350°C)	18.1	15.2	19.6
Bottoms (350°C+)	12.0	14.6	10.5
C3-C4 Yields (wt %)			
C3=	4.8	4.9	4.2
C3	0.6	0.6	0.7
iC4=	1.8	2.1	1.65
nC4=	4.3	4.3	3.45
i C4	2.9	2.9	2.0
n C4	0.50	0.52	0.43
LPG	14.9	15.3	12.5
LPG Olefinicity			
C3 Olefinicity	0.89	0.89	0.85
C4 Olefinicity	0.64	0.66	0.67
iC4=/C4=T	0.30	0.33	0.32
iC4/C4s	0.85	0.85	0.82

reaching higher degrees of isomerization of light olefins (C4 and C5) than the naphthenic VGO. At 70 wt % conversion, iC4=/C4=T ratio is 0.33 with the paraffinic feed, instead of 0.30 with the naphthenic VGO. The value of this ratio corresponding to thermodynamic equilibrium at 530°C is 0.449 (8,9), calculated thermodynamic equilibrium approaches are respectively 73.5% and 67% for paraffinic feed and naphthenic VGO.

The degree of isomerization of C5 olefins is improved in the case of the paraffinic feed: at 75 wt % conversion (see Table 3), iC5=/C5=T ratio reaches 0.60, instead of 0.58 for the naphthenic VGO). This corresponds to calculated thermodynamic equilibrium approaches of 84% and 81% respectively (iC5=/ C5=T ratio corresponding to thermodynamic equilibrium at 530°C = 0.715).

Table 2. Composition and Quality of Gasoline Cut Obtained by Catalytic Cracking of Paraffinic and Naphthenic Feedstocks (MAT Results - T = 530°C)

	Comparison at same conversion level (~75 wt %)		
Nature of the feed	VGO	VGO + Resid	Resid
Character	Naphtenic	Paraffinic	Paraffinic
C/O	8.3	5.7	4.7
Coke (wt %)	6.5	5.5	5.8
Delta-coke	0.79	1.0	1.2
Composition of Gasoline Cut (C5-221°C) (wt %)			
n-paraffins	2.6	3.9	4.5
i-paraffins	19.1	25.4	23.8
Olefins	21.7	25.4	21.7
Naphthenes	6.9	5.7	5.1
Aromatics	49.7	39.6	44.9
RON	95.5	91.5	91.2
MON	83.2	80.2	80.2

Degrees of isomerization of C4 and C5 paraffins are also far higher than the thermodynamic equilibriums at 530°C (respectively 0.34 and 0.68). This shows that degrees of isomerization of C4 and C5 hydrocarbons are fixed by cracking and hydrogen-transfer reactions rather than isomerization reactions.

Table 2 gives the composition and quality of gasoline cuts obtained for both feedstocks at 75 wt % conversion. Paraffinic feed gives a lower aromatics content in the gasoline cut (40 vs 50 wt %), which explains the lower octane value obtained by GC analysis.

FCC operation can be coke-limited by constraints in the regenerator temperature or the air blower capacity. To illustrate this way of operating an FCC unit, comparison of performances obtained with each feed was done at 6 wt % coke yield, which is close to the limit of most industrial units.

At 6 wt % coke yield, the maximum gasoline point is reached on the naphthenic feed, while the overcracking zone is already reached with the paraffinic feed. At this level of coke production (see Table 4), delta-coke is essentially the same for both feeds (0.78), where the same regenerator temperature is expected. Paraffinic feed gives a higher level of conversion (+8 points), especially more LPG (+5.5 points) and more gasoline (+3.3 points), in the same time than a better bottoms upgrading (+2.7 points) and a lower LCO yield (–5.7 points). Significantly higher light olefins are reached with the paraffinic feed: typically +27% rel. for C3= and +45% rel. for C4=.

Table 3. Composition of C5 Cut Obtained by Catalytic Cracking of Paraffinic and Naphthenic Feedstocks (MAT Results - T = 530°C)

	Comparison at same conversion level (~75 wt %)		
Nature of the feed	VGO	VGO + Resid	Resid
Character	Naphtenic	Paraffinic	Paraffinic
C/O	8.3	5.7	4.7
Coke (wt %)	6.5	5.5	5.8
Delta-coke	0.79	1.0	1.2
C3-C4 Yields (wt %)			
C3=	5.6	5.6	5.3
C3	0.7	0.7	0.9
iC4=	2.0	2.5	2.0
nC4=	4.6	5.2	4.3
iC4	3.4	3.4	2.7
nC4	0.6	0.6	0.5
LPG	16.9	18.0	15.7
LPG olefinicity			
C3 olefinicity	0.89	0.89	0.85
C4 olefinicity	0.62	0.66	0.66
iC4=/C4=T	0.30	0.32	0.32
iC4/C4s	0.85	0.85	0.84
C5 Cut yield (wt % feed)	11.71	13.59	9.52
iC5	4.08	4.48	3.00
nC5	0.41	0.53	0.37
C5=T	7.22	8.58	6.15
C5 olefinicity	0.62	0.63	0.65
Degree of isomerization of C5			
iC5=/C5=T	0.58	0.60	0.51
iC5/C5s	0.91	0.89	0.89

These results show that paraffinic feeds, even with essentially the same ConC and hydrogen content, are much easier to crack than naphthenic feeds, which need to be processed at a much higher C/O ratio to reach the same conversion. Selectivities towards olefins production are significantly higher at the same conversion. The selectivities obtained with the paraffinic feed are consistent with a lower importance of hydrogen-transfer reactions than in the case of naphthenic feedstocks: more C4 and C5 olefins, higher degrees of isomerization of butenes and pentenes, lower aromatics content in the gasoline cut, and lower coke yield.

Table 4. Performances and Olefins Yields on Paraffinic and Naphthenic Feedstocks (MAT Results - T = 530°C)

	Comparison at same coke yield (~6.0 wt %)		
Nature of the feed	VGO	VGO + Resid	Resid
Character	Naphtenic	Paraffinic	Paraffinic
C/O	7.7	7.7	5.1
Conversion	71.3	79.7	75.3
Coke (wt %)	6.0	6.0	6.0
Delta-coke	0.78	0.78	1.2
H2 + C1-C2	2.6	2.2	2.5
LPG	16.0	21.5	17.9
Gasoline (C5-221°C)	46.7	50.0	48.9
LCO (221-350°C)	17.6	11.9	16.3
Bottoms (350°C+)	11.1	8.4	8.4
C3-C4 yields (wt %)			
C3=	5.1	6.5	5.8
C3	0.7	0.9	0.9
iC4=	1.9	2.8	2.2
nC4=	4.4	6.4	5.1
i C4	3.3	4.1	3.2
n C4	0.6	0.8	0.7
LPG	16.0	21.5	17.9
LPG olefinicity			
C3 olefinicity	0.88	0.88	0.87
C4 olefinicity	0.62	0.65	0.65
iC4=/C4=T	0.30	0.30	0.30
iC4/C4s	0.85	0.84	0.82

If we consider the case of a coke-limited operation, paraffinic feed gives at the same coke yield much higher conversion and olefins yields than the naphthenic feedstock.

3.2 Evaluation of Real Paraffinic Feedstocks: Resid and Mixture of VGO + Resid

The mixture 90% VGO + 10% Resid (previously compared to the naphthenic VGO) and the pure paraffinic resid both have a pronounced paraffinic character (see Table 5). The pure resid brings higher values for density (0.916

Table 5. Main Characteristics of Naphthenic and Paraffinic FCC Feedstocks

Nature of the feed	VGO	VGO + Resid (90/10 wt %)	Resid
Character	Naphtenic	Paraffinic	Paraffinic
Density 15°C (g/cm³)	0.932	0.8925	0.9159
Con. C (wt %)	0.18	0.89	5.3
Hydrogen (wt %)	12.37	12.59	11.66
Sulphur (wt %)	0.3	0.2	0.2
Nitrogen (wt ppm)	1530	500	1400
Basic nitrogen (wt ppm)	633	233	493
TBP (°C)			
5 wt %	345	323	303
10 wt %	370	355	332
30 wt %	404	412	411
50 wt %	431	437	457
70 wt %	464	470	567
90 wt %	516	553	715
95 wt %	537	695	

versus 0.892) and Conradson Carbon (5.3 versus 0.89), lower hydrogen content (11.7 wt % versus 12.6 wt %), and shifts the distillation curve to higher temperatures.

Table 1 shows the comparison of MAT results obtained with these two feeds at the same conversion level (70 wt % conversion). The paraffinic resid gives more coke and dry gas, and higher LCO yields, to the expense of LPG. The lighter feed (VGO + Resid) is able to give higher productions of olefinic LPG (+17% rel. for C3=, about +25% for C4= and +45% rel. for iC4).

Table 6. Main Characteristics of FCC Catalysts

Catalyst	A	B
Total surface area (m²/g)	158	144
RE2O3 (wt %)	0.8	0
Zeolite unit cell size (Å)	24.24	24.24
Ni (wt ppm)	2170	—
V (wt ppm)	3090	—
Ni + V (wt ppm)	5260	—

Comparison of results at the same coke yield (6 wt %) shows that the overcracking zone is reached on both feeds, allowing the maximization of olefins production (Table 4). The mixture 90% VGO + 10% Resid, compared to the pure paraffinic resid, gives a lower delta-coke value (0.78 instead of 1.2), and can therefore be processed at a higher C/O ratio, allowing a higher conversion level (+4.4 points), higher gasoline yield (+1.1 point), lower dry gas yield (–0.3 point), and higher olefins yield as well (+12% rel. for C3=, and about +30% rel. for butenes).

This comparison shows that, among two paraffinic feedstocks, the lighter one gives lower delta-coke values and coke yields, and higher olefins yields.

3.3   Evaluation of a Pure Paraffinic Feed

The aim of this section is to evaluate the maximum olefinic LPG yields that can be reached by catalytic cracking using the most favorable conditions (100% paraffinic feedstock, choice of the catalyst, higher cracking temperatures). This feedstock (350–530°C) is only composed of long paraffins (C19 to C37 with a maximum centered on C24 (Table 7). Normal/Iso-paraffins balance is 85/15 wt %.

Data given in Table 8 show that the waxy feedstock is very easily cracked: conversions are between 97 and 99.5% (best conversions correspond to highest C/O ratio and temperature). Under these high severity conditions coke yield remains very low, under 2.5 wt %. C7⁻ olefins yields (Table 8) significantly increase at higher temperatures, from 45.4% at 530°C up to 66.8% at 665°C, propylene and butenes being the principal products. Heptenes and higher olefins yields are very low (less than 1%), due to the higher reactivity of these olefins, and decrease when temperature increases. When cracking temperature increases, the yields of C6 and C7 decrease significantly and that of pentenes goes through a maximum, while ethylene yield is increasing drastically.

Olefinicities of C2 and C3 cuts are nearly constant with temperature (Table 9). Ethylene and propylene have little sensitivity to hydrogen transfer

Table 7.   Composition of a 100% Paraffinic Feed (hard wax 48–52°C) by Carbon Number (wt %)

C19	0.06	C24	15.19	C29	3.07	C34	0.32
C20	0.15	C25	14.62	C30	1.68	C35	0.22
C21	1.2	C26	13.03	C31	0.99	C37	0.15
C22	4.94	C27	8.64	C32	0.7	C37	0.07
C23	10.59	C28	5.34	C33	0.45		

Table 8. C1 and Olefins Yield (wt %) Obtained with the 100% Paraffinic Feed (C/O = 5.2) at Different Cracking Temperatures

T	C1	C2=	C3=	C4=	C5=	C6=	C7=	Total
530°C	0.3	0.6	10.2	14	12.9	6.6	0.7	45.4
600°C	0.6	1.5	16	17	11.7	4.6	0.3	51.4
630°C	1.0	2.8	19.6	20.1	13.7	4.3	0.1	60.8
645°C	1.3	3.5	21	21.3	14.1	4.6	0.1	64.7
665°C	2.0	5.6	23.3	21.3	12.5	3.9	0.1	66.8

Table 9. Olefinicity of C2-C7 Cuts Obtained with the 100% Paraffinic Feed (C/O = 5.2) at Different Cracking Temperatures

T	C2 Olefinicity	C3 Olefinicity	C4 Olefinicity	C5 Olefinicity	C6 Olefinicity	C7 Olefinicity
530°C	0.76	0.88	0.62	0.55	0.47	0.11
600°C	0.76	0.90	0.68	0.57	0.44	0.07
630°C	0.77	0.91	0.71	0.62	0.54	0.04
645°C	0.77	0.91	0.75	0.66	0.57	0.04
665°C	0.78	0.91	0.76	0.67	0.68	0.06

due to their low adsorption on catalyst. Olefinicity of C4–C6 cuts increases with temperature, cracking reactions being more favored at higher temperature than hydrogen-transfer reactions. C7 and higher olefins are highly reactive and thus undergo extensive cracking and hydrogen-transfer reactions, leading to a very low olefinicity in the corresponding cuts.

As in the case of paraffinic feedstocks, iC4=/C4=T ratio is constant and equal to 0.33 which is under thermodynamic equilibrium value (about 0.4), while the degree of isomerization of butanes (noted iC4/C4s) is ranging from 0.71 to 0.80, far above thermodynamic equilibrium value (about 0.34).

Dry gas yield increases rapidly, from 1.1 wt % at 530°C up to 9.4 wt % at 665°C, and is mainly due to methane and ethylene production.

Cyclic hydrocarbons, mainly aromatics, are formed from pure paraffins, but aromatics yield is very low and decreases slowly when temperature increases, from 10 wt % at 530°C up to 7 wt % at 665°C. This corresponds to the aromatics content in the gasoline cut of only 15–20 wt %. This is probably why hydrogen-transfer reactions occur only to a limited extent with paraffinic feedstocks.

So, paraffinic feeds appear as ideal feedstocks for olefins production with low dry gas and coke yields. Using a temperature limited to 600°C appears to be the best compromise to maximize C3$^+$ olefins yield, while keeping dry gas yield to acceptable value. At higher temperature the increase of olefins yield is mainly due to ethylene production.

4 CONCLUSION

This study shows that catalytic cracking of a real paraffinic FCC feedstock is much easier to crack than a naphthenic VGO having the same ConC and hydrogen content. Comparison of selectivities at the same conversion shows that processing a paraffinic feedstock rather than a naphthenmic VGO have several beneficial effects on the following parameters: a) improved olefinicity of C4 and C5 cuts; b) improved degrees of isomerization of butenes and pentenes; c) lower aromatics content in the gasoline cut, which is then more olefinic; d) lower coke production. So, it appears that cyclization reactions of paraffins, which are necessary to form aromatics and polyaromatics (coke included), are very slow. This is probably the reason why hydrogen-transfer reactions occur to a limited extent when cracking a paraffinic feedstock. On the contrary, aromatics hydrocarbons are easily formed from naphthenes, and this favors hydrogen-transfer reactions in the case of naphthenic feedstocks.

REFERENCES

1. P.G. White, O.G.J., 1968, 66 (21), 112.
2. E.L. Whittington, J.R. Murphy, I.H. Lutz, Natl. Meet. A.C.S., Div. Petr. Chem., New York, Aug. 27-Sept. 1, 1972, B66-B82.
3. A. Corma, F. Mocholi, V. Orchilles, G.S. Koermer, R.J. Madon, "The Hydrocarbon Chemistry of FCC Naphtha Formation," Ed. H.J. Lovink and L.A. Pine, ACS Symp., Div. Petr. Chem., Sept. 10-15, 1989, 19.
4. P. Venuto, E.T. Jr. Habib, *Fluid Catalytic Cracking with Zeolite Catalysts*, Marcel Dekker, New York, 1979.
5. J. Abbott, B.W. Wojciechowski, *J. Catal., 107*, 571 (1987).
6. A. Corma, A.L. Agudo, *React. Kinet. Catal. Lett., 16*, 253 (1981).
7. D. Rawlence, "Factors Influencing LPG and Gasoline Composition," Third Intercat Symposium on FCC Additives, March 4, 1994, Tokyo (Japan).
8. T.G. Roberie, G.W. Young, D.S. Chin, R.F. Wormsbecher and E.T. Habib, "Reformulated Gasoline: Current and Emerging FCC Catalysts," 1992 NPRA Annual Meeting, New Orleans, March 22–24, 1992.
9. J.B. McLean, G.S. Koermer and R.J. Madon, "Reformulated Gasoline Catalyst's Impact on FCCU," 1992 NPRA Annual Meeting, New Orleans, March 22–24, 1992.

RECENT ADVANCES IN FCC CATALYST EVALUATIONS: MAT VS. DCR PILOT PLANT RESULTS

Lori T. Boock and Xinjin Zhao
GRACE Davison
Division of W. R. Grace & Co.-Conn.
7500 Grace Drive, Columbia, MD 21044
(410) 531-4173

Recent work on laboratory catalyst evaluations in the Davison microactivity unit (MAT) and the Davison Circulating Riser (DCR) have shown that there are differences in catalyst performance and rankings in the two units. These differences, however, a re clearly understandable in terms of catalyst properties, feedstock properties and reactor contact times. Due to the longer catalyst contact time in the MAT unit, catalysts with a higher active matrix component tend to produce more coke and give overall poorer yields in the MAT unit than the DCR. Correspondingly, high zeolite catalysts perform much better in the MAT unit, relative to the higher matrix catalysts. These differences tend to be exaggerated when heavy feedstocks are used. These results suggest that while both the MAT unit and the DCR unit can be used to rank catalyst performance, one must be careful when comparing catalysts with very different purposes (i.e., bottoms cracking catalysts vs. octane catalysts) and formulations.

I. Introduction and Background

One of the challenges in evaluating new FCC catalyst technologies has been in simulating how the catalyst will perform in a commercial FCC unit. Commercial FCC performance data is often valuable, but commercial FCCs are never started up on a single, fresh FCC catalyst, feedstocks and FCC operating parameters often vary over a short time, and commercial FCC data is sometimes difficult to interpret. Two common laboratory techniques used to evaluate the performance of FCC catalysts include microactivity testing (1) and circulating riser pilot plants (2). Both units are able to evaluate both equilibrium catalysts and laboratory deactivated catalysts. Typical results include catalyst activity and yields. Each type of laboratory procedure has its own set of advantages and disadvantages. Table I summarizes some of the differences between the MAT and DCR units.

The standard Microactivity (MAT) test was developed in the 1960's and a standardized procedure has been published as ASTM D 3907-87. The Davison MAT unit, which incorporates modifications to the ASTM procedure, is a quick and easy to operate procedure to measure the activity and selectivity of equilibrium and laboratory deactivated FCC catalysts. The Davison MAT requires only a small amount of catalyst and feed, gives excellent reproducibility and can process a wide range of feedstocks. MAT testing is the workhorse protocol routinely used at Davison and many other laboratories for catalyst evaluation.

The Davison Circulating Riser (DCR) was developed in its present form in 1986. The DCR operates as a fully circulating pilot unit and can be run in both adiabatic (standard operation) and isothermal mode. The DCR is more complex to operate than the MAT and testing catalysts in the DCR unit is more time consuming, requires more catalyst and feed, and is more expensive; however, the DCR does operate more like a commercial FCC unit, and yield structures are similar (3). Since many laboratories use both types of tests, the question arises if both units always give the same results, and if not, which results are accurate? This paper seeks to answer these questions by exploring when DCR and MAT results are different, and by examining some of the factors which may explain these differences.

II. Catalyst Effects

A number of catalyst evaluations have been performed in both the DCR unit and the MAT unit, using the same catalysts and feedstocks. These tests have shown that in most cases, the catalyst rankings and yield differences were the same or very similar; however, in a few cases, catalyst rankings were reversed. Table I summarizes these tests in terms of catalyst, feedstock and deactivation types.

Some key observations can be made from Table II. In all cases, at constant conversion, the DCR produced more olefins (both gasoline and light gas) and less hydrogen than the MAT unit (hydrogen difference was smallest for high Ni catalysts). These results are likely due to the temperature and severity differences between the two units. However, gasoline yields were lower and LCO and coke yields were higher in the MAT than DCR for active matrix catalysts, and the reverse was true for inert matrix catalysts. Bottoms yields were usually similar, but the differences between the relative bottoms cracking of catalysts varied. The change in coke and gasoline yields can be partially attributed to contact time differences, as discussed below. In the cases where ranking reversals were observed, the catalysts being compared were always of different types, i.e., an active matrix catalysts vs. a inert matrix catalyst. Additionally, the ranking reversals followed a clear pattern: active matrix catalysts tested better in the DCR than inert matrix catalysts and vice-versa in the MAT unit, particularly with respect to coke and gasoline selectivities.

As described above, rankings for coke yields can sometimes be quite different. Estimated coking rates of zeolite and matrix catalysts (determined from a MAT test with varying contact times) are significantly different as shown in Figure 1. Inert matrix catalysts (all zeolite cracking) initially coke up rapidly, as

Table I
Typical Differences Between MAT and DCR

MAT	DCR
» longer contact time	» short contact time
» fixed bed	» fluidized bed
» reactor T set at bottom of reactor - 50°F temperature drop across bed	» adiabatic - reactor setpoint at top of reactor
» feed contacts fresh catalyst	» catalyst well mixed when contacts feed
use of glass beads	no glass beads

Table II

Summary of DCR vs. MAT Catalyst Evaluations

Catalyst Type	Deactivation Type	Ranking Change	Selectivity change - MAT vs. DCR
Inert matrix, varying sieve content	4@1500°F	no change	At constant conversion, DCR produced less hydrogen, more olefins, less gasoline, more coke and LCO
Medium to high active matrix catalysts	4@1500°F	no change	At constant conversion, DCR produced less hydrogen, more olefins, more gasoline, less coke and LCO
Inert vs. active matrix	4@1500°F	ranking change	At constant conversion, DCR produced less hydrogen, more olefins. Al-sol produced more coke in MAT, but at higher activity.
Bottoms cracking vs. standard	CPS	ranking change	DCR produced less hydrogen, more olefins. Bottoms cracking catalyst gave better yields in DCR
Bottoms cracking vs. standard	CPS (high Ni)	ranking change	DCR produced less olefins. Bottoms cracking catalyst gave better yields in DCR than si-sol, reverse in MAT
Varying degrees of active matrix	CPS	no change	Differences in yields were smaller in DCR. DCR produced less hydrogen and more olefins
Inert vs. active matrix	CPS	ranking change	Si-sol looked significantly better in MAT, Al-sol in DCR

Coking Rate of Zeolite and Matrix Catalysts

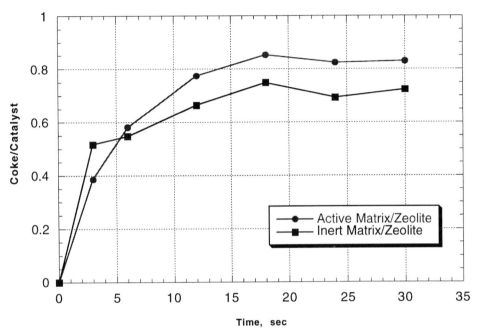

Figure 1

highly active sites in the small pores are blocked; further coke build-up is negligible. On the other hand, coke on active matrix catalysts builds up slower, but continues to increase. Typical MAT testing is conducted at a contact time of 30 seconds, while DCR testing is conducted at a contact time of 4-5 seconds, which is similar to most commercial operations. The differences in coke yields noted above (and likely LCO and gasoline yields) can be explained by these contact time differences. Thus, any pilot plant unit with longer contact times may unfairly penalize active matrix catalysts.

III. Feedstock Effects

Other than the differences in contact time between MAT and DCR, we have also observed that feed properties also play a significant roles in the differences between MAT and DCR yields. We have observed more cases of coke yield ranking reversals between catalysts with active matrices and inert matrices, especially with feeds contains a significant amount of Conradson carbon (e.g. >4 wt%). Part of the reason for the coke ranking reversal might be related to the differences in preheating systems between the MAT and DCR. The preheating system and catalyst/feed contact area in a DCR is similar to a commercial FCC unit, whereas in a MAT unit, since the catalyst bed is fixed and at a lower temperature, the feed/catalyst contact is very different. If a preheating system is precracking the feed before the feed contacts the catalyst bed, the feed that the catalyst actually contacts would be somewhat different from the original feed. The net effects on catalyst due to the feed property changes may or may not be the same for different catalysts. The preheating of the fixed catalyst bed temperature. In most MAT designs, however, for heavy feeds, this goal may not be completely achieved and feed precracking is likely. system in the MAT should be designed in a way that precracking is minimized while a high enough temperature is achieved to maintain the isothermal control

Figures 2 and 3 show the yield differences between catalysts with active and inert matrices in the MAT and DCR. MAT typically gives the

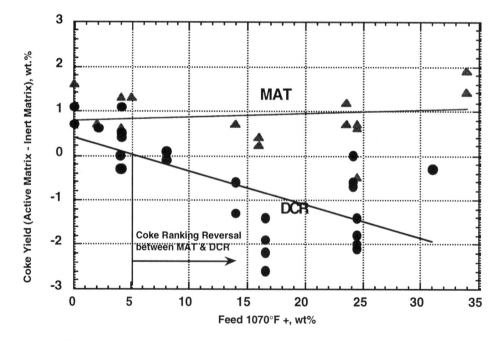

Figure 2
Coke Rankings - MAT vs. DCR

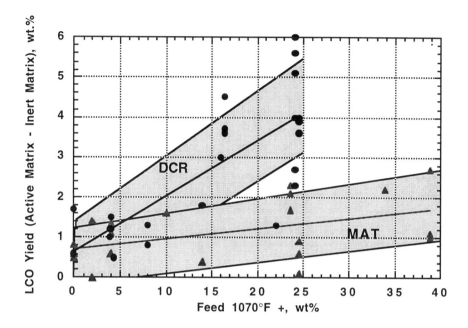

Figure 3
Bottoms Cracking Rankings - MAT vs. DCR

same ranking as the DCR for relatively light feeds. As we mentioned before, coke rankings can be reversed for heavy feeds.   On the other hand, MAT and DCR typically give the same ranking for bottoms conversion, though the DCR differentiate catalysts substantially better than the MAT, especially for heavy resid feeds.

IV. A Typical Catalyst Evaluation

Results are shown in Tables III-V for a typical catalyst evaluation where a dramatic ranking reversal was observed. The catalysts tested were a selective-active matrix catalyst, designed for its metals tolerance and bottoms cracking, and a metals tolerant catalyst with very little active matrix. Catalyst properties are shown in Table III. The catalysts were deactivated by Davison's CPS (4) procedure and run in the MAT unit and DCR unit with a heavy Pacific feedstock. The catalysts were also run in the DCR with a number of different feedstocks, as shown in Table IV. Table V compares the activity and yields.

Table III
A Typical Laboratory Evaluation
Catalyst Properties

			Inert Matrix	Active Matrix
CHEMICAL ANALYSES:				
Al2O3	:	Wt.%	35.1	48.2
RE2O3	:	Wt.%	1.80	1.45
Na2O	:	Wt.%	0.38	0.28
SO4	:	Wt.%	0.89	0.26
PHYSICAL ANALYSES:				
2 @ 1100 F				
ABD			0.81	0.73
DI			6	6
SA	:	M2/GM	300	305
Zeolite	:	M2/GM	253	240
Matrix	:	M2/GM	47	65
CPS:				
SA	:	M2/GM	213	201
Zeolite	:	M2/GM	182	˙166
Matrix	:	M2/GM	31	35
Unit Cell			24.28	24.29
Ni	:	ppm	2702	2815
V	:	ppm	2290	2560
MA	:	Wt.%	73	68
H2	:	SCFB	254	282
C	:	Wt.%	5.8	5.7

Table IV
A Typical Laboratory Evaluation
Feedstock Properties

Feedstock:	Std. VGO	Light Arabian	Heavy Pacific	Std. Resid
Feed Number:	F92-444	F94-245	F94-513	F90-302
API @60 F	25.8	20.2	23.0	25.6
Aniline Pt. F	189	193	199	240
Sulfur wt%	0.301	2.77	0.38	0.31
Total Nitrogen wt%	0.12	0.08	0.09	0.19
Basic Nitrogen	0.052	0.02	0.03	0.06
ConCarbon wt%	0.53	1.23	3.61	3.98
SIMDIST Distillation: Vol%, Temp °F				
ibp	252	495	522	415
10	568	689	687	662
20	655	734	741	752
40	754	804	819	891
60	828	874	907	1054
80	923	952	1040	1236
95	1040	1044	1333	1357
end	1202	1171	-	-
'K' Factor	11.78	11.65	11.93	12.39

It is clear from Table V that the performance of the two catalysts is quite different in the MAT vs. DCR. In the MAT unit, the metals tolerant catalyst with no active matrix is the clear winner in both activity and selectivity. However, the reverse is true in the DCR with the heavy Pacific feed. The selective active matrix catalyst also gives better yields (primarily coke and gasoline) in the DCR than the catalyst with inert matrix on all the feeds tested and the difference in yields is larger for the heavier feeds. These yield differences are consistent with the shorter contact time of the DCR.

Table V
A Typical Laboratory Evaluation
Yields at Constant Conversion

Inert Matrix		MAT Heavy Pacific	DCR Heavy Pacific	DCR Std. Resid	DCR Std. VGO	DCR Light Arabian
Conversion	: wt%	77	75	81	73	73
C/O		3.5	5.3	6.3	7.3	6.1
H2	: wt%	0.38	0.09	0.08	0.06	0.14
Gasoline	: wt%	54.7	52.1	50.5	51.3	46.5
LCO	: wt%	17.6	15.5	12.4	18.1	16.6
Bottoms	: wt%	5.4	9.5	6.6	8.9	10.4
Coke	: wt%	5.8	5.5	8.3	4.1	6.6

Active Matrix		MAT Heavy Pacific	DCR Heavy Pacific	DCR Std. Resid	DCR Std. VGO	DCR Light Arabian
Conversion	: wt%	77	75	81	73	73
C/O		4.2	3.0	6.8	7.5	6.1
H2	: wt%	0.49	0.10	0.10	0.08	0.21
Gasoline	: wt%	52.4	55.3	52.0	53.1	48.4
LCO	: wt%	18.0	17.3	13.9	19.6	17.6
Bottoms	: wt%	5.0	7.7	5.1	7.4	9.4
Coke	: wt%	6.5	4.2	7.7	3.8	6.3

V. Summary and Conclusion

Zeolite and matrix cracking have different coking rates associated with them. This results in different yield structures when catalysts with a high active matrix content are compared to high zeolite/inert matrix catalysts in short contact time reactors, such as the DCR, versus longer contact time units, such as MAT. The short contact time of the DCR is more consistent with typical commercial FCC units. These results are exaggerated when heavy feedstocks are used, since in the DCR, the catalyst/feed contact is similar to a commercial unit, whereas in the MAT it is quite different and some feed precracking may occur, especially with heavy feeds. This result suggests that care must be taken when catalysts with very different purposes are evaluated in the MAT. However, MAT studies are extremely valuable as a quick and inexpensive "workhorse" tool for evaluating catalyst performance and discerning yield differences, especially if catalysts with similar objectives are being compared.

These results also suggests that there is value in developing a shorter contact time MAT unit and in modifying the way in which the MAT processes heavy feeds. These projects are currently underway in Grace Davison's research laboratories.

References:
(1)   Moorehead, E.L., J.B. McLean and W.A. Cronkright, "Microactivity Evaluation of FCC Catalysts in the Laboratory: Principles, Approaches and Applications" in   J.S. Magee and M.M. Mitchell, Jr., Fluid Catalytic Cracking: Science and Technology, Studies in Surface Science and Catalysts, Vol. 76, 1993. Elsevier Science Publishers, Ch. 7.
(2)   Young, G.W. and G.D. Weatherbee. "FCCU Studies with an Adiabatic Circulating Pilot Unit." AIChE Annual Meeting in San Francisco, CA 1989.
(3)   Young, G.W., G.D. Weatherbee and S.W. Davey. "Simulating Commercial FCCU Yields with the Davison Circulating Riser Pilot Unit." NPRA Annual Meeting in San Antonio, TX, 1988.
(4)   Boock, L.T., T.F. Petti and J.A. Rudesill. "Recent Advances in Contaminant Metal Deactivation and Metal Dehydrogenation Effects During Cyclic Propylene Steaming of FCC Catalysts." International Symposium on the Deactivation and Testing of Hydrocarbon Conversion Catalysts in Chicago, IL 1995.

COKING AND CHEMICAL STRIPPING OF
FCC CATALYSTS IN A MAT REACTOR

J.R. Bernard, P. Rivault, D. Nevicato, I. Pitault*, M. Forissier*, S. Collet
Elf Aquitaine, CRES, B.P. 22, 69360 Solaize, France.
*LGPC/CNRS-CPE LYON, B.P. 2077, 69616 Villeurbanne, France

The knowledge of coking and stripping is a key point to understand heat balance and associated phenomena of the FCC plant. Coking kinetics was studied in the Micro Activity Test (MAT) by working at various C/O (up to 60) and various time on stream with three different VGO. It appears that a front of constant coke content develops in the bed when C/O is decreased. This confirms the rapid coking essentially from the feedstock.

On the other hand, the influence of stripping duration after oil injection is studied between 1 and 15 min. During this time interval, up to 20% of the coke can be converted to fuel gas, depending on cracking pressure and temperature. This amount depends also strongly on Nickel and Vanadium on the catalyst. This shows that the coke is subjected to cracking reactions in a FCC stripper.

Fuel gas yields are strictly proportional to coke yields after extensive stripping, even at high conversion and various temperature; this proves that fuel gas is produced only by thermal cracking and coke chemical stripping.

1. Introduction

The behavior of the FCC unit is determined by its heat balance requirement, since exothermic heat of coke combustion in the regenerator provides endothermic heat of cracking through catalyst circulation.

Two phenomena are important to understand how FCC plants are working: the kinetic of coke formation and spent catalyst stripping.

The kinetic of catalyst fouling by coke is governing the delta coke, i.e. the coke percentage deposited on the catalyst in the reactor and burned in the regenerator, and it interacts on the kinetic constants of cracking reactions (Forissier and Bernard, 1991). The kinetics of coking has been studied in pulse or continuous microreactors with light feedstocks (Lin et al., 1983, Hatcher, 1985, Dadybujor and Liu, 1992, Beinaert et al., 1994). However, these reactors are not suitable when vacuum gas oil or heavier feedstocks are converted.

Turlier et al. (1994) showed that coke from vacuum gas oil is formed extremely rapidly at the bottom of commercial FCC risers. It is composed in part of alkylated polyaromatic hydrocarbons (3 to 7 nuclei) which seem to be trapped in the zeolite structure:

The chemical analysis shows that this coke is similar to coke from light alkanes in Y zeolite (Magnoux, 1987), suggesting that coking chemistry depends more on the zeolite properties than on feedstock analysis. Pregraphitized carbon is also found on the catalyst. Its proportion depends strongly on riser temperature and catalyst time on stream: it is small at low severity (505°C riser outlet temperature) but it increases to 50% at 530°C and up to 90% after 15 min stripping (Turlier et al 1994). This proves that coke is chemically reactive even if it is formed quasi instantaneously on the catalyst.

The coke yield depends on the characteristics of catalyst (acid sites and pore structure), on the nature of the feedstock and on the operating conditions. It is more important for polyaromatic and aromatic feedstocks compared to naphthenic and paraffinic feedstocks (Appleby et al., 1962). The olefinic compounds play an important role in coking reactions by bimolecular reactions (condensation and hydrogen transfer). Experimental works show that coking reactions are primary for catalytic cracking of aromatic and olefinic feedstocks (Appleby et al., 1962, and Nace et al., 1971) contrarily to paraffinic feedstocks (John and Wojciechowski, 1975). The amount of coke in the catalyst is not modified by the variation of the reaction temperature between 480 and 540°C. However, higher temperatures increase the coke probably because of thermal cracking to unsaturated hydrocarbons (Biaou et al 1993, Turlier et al 1994)

If coke is build in the riser, it is modified in the stripper (Upson et al, 1993, Turlier et al., 1994, Venuto and Habib, 1979) where catalyst residence time is in the range of several tens of seconds to several minutes.

The stripper plays a role of coke yield regulator: the catalyst circulation is set by the stripping efficiency to maintain the necessary coke yield to heat balance the plant. Gerritsen et al. (1991) show that at 500°C, fluidized catalyst coke content decreases from 1.5 %wt to 0.7 %wt when stripping time increases from 10s to 200 s. Turlier et al. (1994) observe that the catalyst carbon content does not depend on riser elevation but it decreases significantly after laboratory stripping, while its hydrogen content decreases too via graphitization.

A more efficient stripper allows to work with higher catalyst circulation since the catalyst carries less specific combustion enthalpy to the regenerator. Then the conversion is larger and more selective because of larger catalyst hold up in the riser, less coked catalyst (Forissier and Bernard, 1991) and cooler riser bottom minimizing thermal cracking.

Thus coking and stripping are key points for plant performances and the purpose of this paper is to describe coke formation kinetics and chemical reactions which occur during stripping. Commercial feedstocks and equilibrium catalysts are used in a MAT reactor. Effects of pressure, temperature and stripping duration time are studied.

2. Coking in a MAT reactor.

The reactor is a modified MAT already described (Forissier and Bernard, 1991). Three typical feedstocks (paraffinic Montmirail, naphtenic Nigeria and aromatic Aramco) were used. Their detailed analysis and the catalyst characteristics are given by Pitault et al. (1994).

The coking reactions for vacuum gas oils were carried out on a commercial equilibrium catalyst over a wide range of reaction conditions (table 1). After VGO cracking, and 15 min stripping by inert gas, the catalyst was discharged and well mixed. The average coke mass in the bed was determined by analysis of carbon with a LECO Carbon Determinator CR12.

In figure 1, the coke accumulated in the MAT is shown versus catalyst mass and time on stream (TOS) at 530 °C and at constant feedstock flow rate. Note that the C/O - i.e. the mass ratio of catalyst to the oil fed to the reactor - is inversely proportional to TOS in these experiments. It varies from 1 to 60 at 5s TOS and 6g catalyst. At high TOS (50s.) the coke yield is proportional to the catalyst mass or to the C/O, which means that the final coke content is constant over the bed, but this is not the case with smaller TOS (larger C/O).

To better comment these results, it is worthwhile to derive the curves of figure 1 with respect to the catalyst mass. This yields the local coke content on the catalyst

Table 1. Experimental conditions of the modified MAT.

feedstocks	Aromatic VGO Aramco(0.84%CC 2.75%S)
	Naphtenic VGO Nigeria (0.60 %CC 0.32%S)
	Paraffinic VGO Montmira (0.45%CCR, 0.28%S)
feedstock flow rate	0.02 g/s
injection time (Time On Stream)	5 to 100 s
N_2 stripping gas flow rate	40 cc/min during 15 mn
pressure	1 atm.
temperature	480, 530, 580 and 630 °C
1 to 6 g equilibrium catalyst from Grace Davison, inert matrix, ASTM activity 69, 1250 ppmV, 850ppm Ni dilution with inert quartz to maintain a constant delta P	

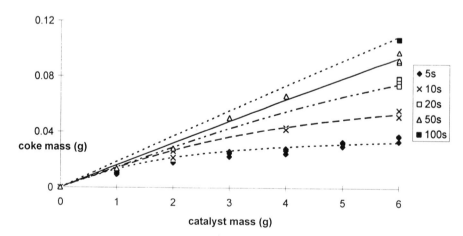

Figure 1. Mass of coke deposited in the catalytic bed as a function of the catalyst mass, Aromatic VGO, T= 530°C, feedstock flow rate=0.02 g/s,

along the bed at various TOS, as shown in figure 2. This content tends to be constant for TOS larger than 20s. This means that coking reactions are rapid and they are then strongly self inhibited as observed by Lopes et al. for n heptane transformation. However there is no real saturation level as shown by the content increase between 20 and 100s TOS. The additional fresh feedstock supplied to the bed during the last tens of seconds contains coke precursors which deposits slowly and evenly on the already coked catalyst. Inversely coke formation from converted feedstock is negligible if sites are already fouled. Turlier et al. (1994) showed that coke itself can disappear in part by cracking to gaseous products and can migrate within the catalyst. This phenomenon occurs certainly also during VGO cracking although it can be detected only during stripping. This would allow more place for coke deposition at large TOS. Moreover the coking properties of the matrix may interfere even if it is given as inert for commercial conditions.

Inversely when TOS decreases from 50 to 5s, coke precursors are lacking, except on the first gram of catalyst whose coke content is close to the maximum. It decreases then significantly along the bed especially at 5s TOS because the last layers of catalyst see small amounts of heavily converted species. For example the coke yield from the aromatic feedstock reaches 37% at C/O 60 (5s TOS, 6g catalyst), suggesting an extremely high conversion, though not measurable.

These observations are consistent with the delta coke profile measured by Turlier et al. (1994) in a commercial riser, where the whole coke is made immediately at the bottom. This is because of the extremely rapid coking kinetic of the feedstock, but when hydrocarbons rise with increased conversion, their coking activity decreases and self inhibition by coke on catalyst occurs. Moreover reaction temperature decreases because of endothermicity so that coking reactions are more inhibited.

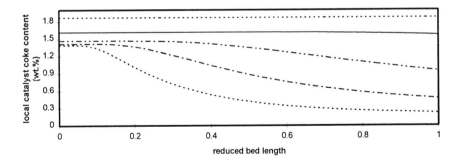

Figure 2. Simulated delta coke profile in the bed, Aromatic VGO, T= 530°C, feedstock flow rate=0.02 g/s, (- -) tos=5s, (-··) tos=10s, (- ··) tos=20s, (—) tos=50s, (---), tos=100 s.

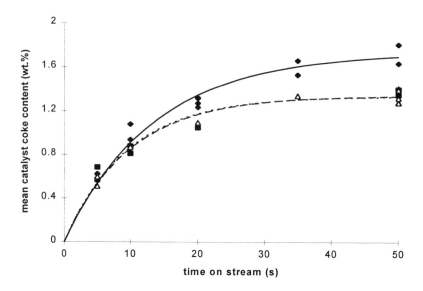

Figure 3. Mean catalyst coke content in the catalytic bed as a function of the time on stream for various VGO. T= 530°C, feedstock flow rate=0.02 g/s, catalyst mass=6g. (◆) Aromatic VGO, (Δ) Naphtenic VGO ,(■) Paraffinic VGO.

Figure 3 shows the influence of the feedstocks on the coke accumulated in the reactor. Aromatic VGO (Aramco) gives more coke than naphtenic and paraffinic VGO (Nigeria and Montmirail). Results similar to those shown in figure 2 were obtained with these two last gas oils.

The cracking temperature does not change significantly the mean delta coke when it is lower than 580°C or when TOS is smaller than 20s (figure 4). Beyond these limits, the increase is due probably to an increased coke precursors concentration from feedstock thermal cracking (coke from olefins condensation).

It is striking to observe that at 5s TOS, a temperature variation from 480 to 630°C does not change the mean delta coke although it is not known whether the coke profile in the bed is altered. The conversion is likely very high at every temperature and thermal cracking at high temperature and high conversion cannot produce efficient coke precursors.

3. Stripping in a MAT reactor.

The coke evolution during stripping is also studied in the MAT reactor by varying the stripping duration from 1 to 15 minutes, at the same pressure and temperature

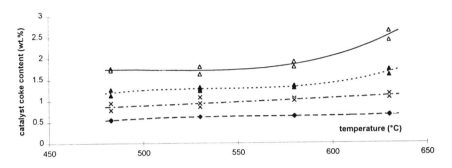

Figure 4. Mean catalyst coke content versus reaction temperature, catalyst mass=6g, Aromatic VGO, feedstock flow rate=0.02g/s, (♦,--) tos=5 s, (X, _._.) tos=10 s, (•, ...) tos=20s, (Δ,___) tos=50 s.

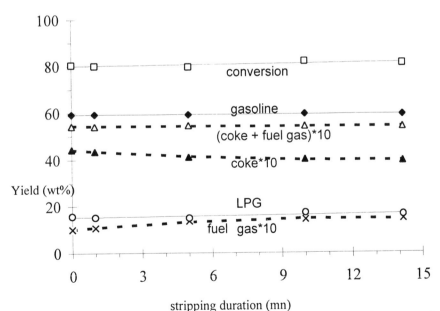

Figure 5. Effluent Yield versus stripping duration, equilibrium catalyst mass=6g, temperature= 480°C ASTM feedstock flow rate=0.02g/s, nitrogen flow rate= 40 cm3/mn NTP, (□) MAT conversion, (♦) gasoline, (Δ) (fuel gas + coke)*10, (▲) coke * 10, (O) liquefied petroleum gas, (X) fuel gas * 10.

than the cracking itself. The equilibrium catalyst is a Super D from Crosfield, with 180 ppm Ni and 500 ppm V. Standard conditions are T = 483 °C, atmospheric pressure, C/O 6, time on steam 50s, nitrogen flow rate 40 cc/min maintained constant during cracking and stripping. As in the standard procedure, cumulative yields are determined, but the good reproducibility allows to detect small yield variations due to changes of stripping time.

Preliminary experiments with various stripping nitrogen flow rates show that the whole bed is completely flushed and desorbed within one minute, and only chemical stripping i.e. coke cracking is occurring after this delay: figure 5 shows that this chemical stripping converts ca 10 % of coke to fuel gas. As the sum fuel gas + coke remains constant, it can be concluded that the production of other lumps from coke is negligible during this stripping step.

Figure 6 shows the variations of light gases yields. Relative variations are ranked as follows: H2 >> CH_4 > C_2H_6, showing dehydrogenation and desalkylation of coke. No significant ethylene variation was detected during stripping.

The same trend appears at 525°C instead of 483°C (figure 6). The relative yield variations seem however smaller. The primary coke is less hydrogen rich. Very probably chemical stripping is also quicker. Its occurrence is more superposed

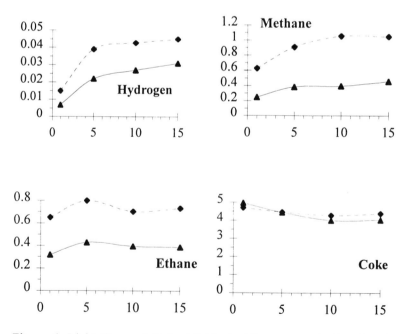

Figure 6. Light Gas and Coke Yields (wt%) versus stripping duration (mn) , ASTM feedstock, TOS 50s, C/O 6, (---- ◆) 525°C and (——▲) 480°C.

with the oil injection. It is consequently less detected during the stripping step. Some experiments were done at 3 bars g. and 483°C; the same trends were detected than at 525°C, with a larger coke production, as already shown (Turlier et al. 1994). Coke analysis is based on carbon content. However figures 5, 6, 7 show that chemical stripping occurs by loss of hydrogen rich gases. Therefore the carbon reduction is accompanied by a strong H/C decrease which plays an important role in the plant heat balance.

It seems that an atomic H/C ratio in the order of 0.4 can be reached after extensive stripping. However a stripping duration of 15 min would be equivalent to a stripper 50% larger than the regenerator of the commercial plant.

In reality, catalyst residence time exceeds rarely 2 min in the largest strippers, after ca 5 to 10s spent in the riser terminated with a quick disengager. The chemical stripping is more intense during the first minutes of stripping (figures 6, 7) and obviously it is occurring also during the 30 or 50s of feedstock injection, and even during the first seconds in the riser. This shows that this phenomenon is important in commercial strippers. Samplings above the stripper of a commercial plant equipped with a quick disengager showed the presence of large amounts of fuel gas and L.P.G.

4. Influence of Nickel and Vanadium on chemical stripping

It is well known that Ni and V produce fuel gas. Specific experiments were done to explore their influence on chemical stripping, by using a catalyst Quantum 2000 from Crosfield. It is used after steaming the fresh sample or after impregnation of each metal. A cyclic method is used to obtain metals effects closer to those observed in commercial plants. A content of 1000 ppm Nickel is reached, and three times more for Vanadium to obtain a comparable effect. Experimental conditions are summarized in Table 2.

Results are shown in figure 7. In these conditions, it is necessary to account for LPG production from coke since the sum LPG + FG + coke varies much less than FG + coke. This difference with the former tests on super D is not clearly understood. It may be attributed in part to shorter TOS (30s instead of 50s) revealing more LPG production during the early seconds of stripping. Nickel does not produce more coke than only steamed catalysts, but this coke is more refractory to chemical stripping: the primary coke is more aromatic and/or chemical stripping is readily catalyzed by Nickel during the oil injection step. Hydrogen production is also strongly enhanced by Ni during stripping, but it contributes little in weight to coke decrease. The same remarks on H/C apply.

Vanadium's behavior is intermediate. Its stripping activity is not negligible even if its concentration is three times larger.

These results show that light gas production from metals can be an important part

Table 2. Experimental conditions for the study of V and Ni impurities.

catalyst	fresh Quantum 2000 from Crosfield
metal deposition	in a fluidized bed reactor with cycles of -deposition (80ppm Ni or V) -regeneration by air
steaming	17 hours at 790°C (without metal) at 730°C (with metal)
reaction and stripping conditions	480°C and 525°C, 6g of catalyst, 1.5g feedstock injected in 30s, Nitrogen flow rate 30cm^3/mn NTP
feedstock	ASTM MAT

of coke chemical stripping. The light gas produced by metals is harmful for downstream compression, but this production contributes to a better stripping with the benefits described earlier.

5. Origin of fuel gas production

Since fuel gas is tight in part to coke, it is worthwhile to draw the correlation. It is in figure 8, where the FG yields are plotted for three different feedstock versus coke yield in MAT. These experiments are those realized in the coke production study, by changing the catalyst hold up at 50s TOS and various temperatures.
Parallel straights are obtained at different temperatures. They do not extrapolate to zero at zero coke yield (which corresponds to zero catalyst). These amounts are of thermal origin. They increase significantly with temperature.
When increasing the catalyst hold up and the proportional coke yield (Fig. 1), the fuel gas increases also proportionally with the same rate at every temperature. Note that when doing these tests, the conversion is not proportional at all and it levels to very high values. This shows that coke built on the catalyst yields always the same fuel gas content after extensive stripping (15 min). This content does not

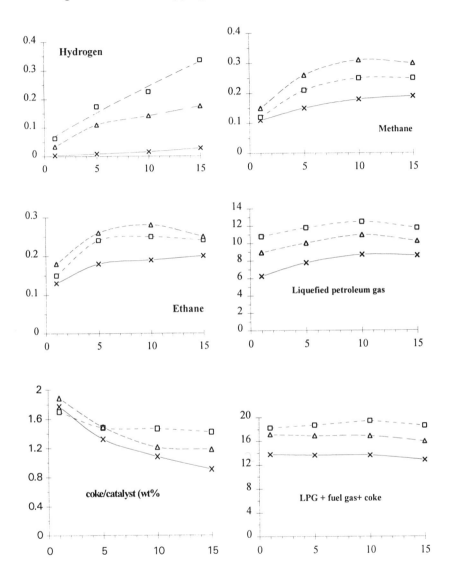

Figure 7. Yields (%wt) versus stripping duration (mn), equilibrium catalyst mass=6g, temperature= 480°C ASTM feedstock flow rate=0.02g/s, nitrogen flow rate= 40 cm3/mn NTP, (X) steamed, (□) Ni 1070ppm, (Δ) V 2920ppm.

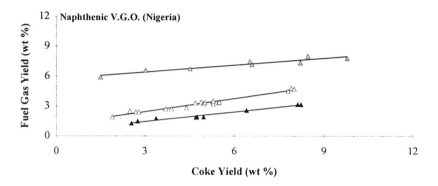

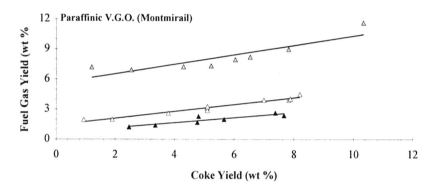

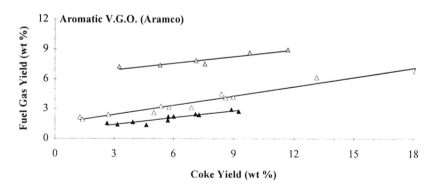

Figure 8. Fuel gas yield (wt %) versus coke yield (wt %), feedstock flow rate = 0.02g/s, nitrogen flow rate= 40 cm³/mn NTP, equilibrium catalyst from Grace Davison and (♦) 480°C, (Δ) 530°C, (▲) 580°C reaction temperatures.

depend on the temperature, but very probably coke cracking is quicker at higher temperature.

Thus fuel gas is produced exclusively by thermal cracking and coke chemical stripping. Coke dehydrogenation to form an aromatic from an olefinic naphthene and coke desalkylation by hydrogenolysis of small chains may explain these results.

6. Conclusion

Coke and fuel gas production are completely decoupled from other cracking reactions. The coke deposition from VGO on FCC catalysts is an extremely rapid phenomenon, even at 483-530°C, when local conversion is small. The production of coke self inhibits strongly beyond a certain level depending of operating conditions (mainly feedstock analysis and pressure). It is governed by the availability of coke precursors and it does not depends on temperature, except if this latter is high enough to add coke precursors by thermal cracking.

When extrapolating to risers where catalyst and feedstock are mixed at the bottom, this confirms that the whole coke is made immediately.

The coke is then chemically changed by stripping: part of it is cracked to fuel gas ($H_2CH_4.C_2H_6$) and LPG, contributing to the overall stripping which was formerly seen only as a sweeping-desorption operation. Not only the delta coke may decrease by a maximum of 10-20% wt for VGO, but its H/C decreases also significantly.

Nickel and Vanadium have a strong influence on this last phenomenon. Their detrimental effect on light gas production occurs mainly in the stripper.

Fuel gas is produced exclusively from thermal cracking and coke chemical stripping. The amount of fuel gas produced from extensive coke stripping does not depend on temperature, but only on coke yield, for given catalyst and feedstock.

When moving to the commercial operation, one could extrapolate that increasing the stripper volume could rise the light gas yield. This is always the contrary, because the overall stripping efficiency increases (drop of delta coke and its H/C). Then catalyst circulation increases, regenerator temperature decreases and thermal production of fuel gas diminishes, this last effect being more important than the coke cracking effect.

7. References

Appleby W.G., Gibson J.W. and Good G.M., 1962, Coke formation in catalytic cracking, Ind. Eng. Chem. Proc. Des. Dev., 1, 102-110.

Beirnaert H.C., Vermeulen R. and Froment G.F., 1994, A recycle electrobalance reactor for the study of catalyst deactivation by coke formation, Catalyst deactivation 1994, Studies in Surface Science and Catalysis, ed by Delmon B., Froment G.F., Elsevier Science, Amsterdam, 97-112.

Biaou D., Cartraud P., Magnoux P. and Guisnet M., 1993, Coking, aging and regeneration of zeolites. XIV Kinetic study of the formation of coke from propene, Appl. Catal. A, 101, 351-369.

Dadyburjor D.B. and Liu Z., 1992, coking in pulse and flow microreactors, *Chem. Eng. Sci.*, **47**, 645-651.

Forissier, M. and Bernard, J.R., 1991, Deactivation of cracking catalyst with vacuum gas oil, Catalysts deactivation 1991, C.H. Bartholomew, J.B. Butt editors, Elsevier, Amsterdam, 359-366.

Gerritsen L. A., Winjgards H. N., Verwoert J. and O'Connor P., 1991 Akzo Catalyst symposium Fluid catalytic cracking, Scheveningen, 123.

Hatcher W.J., 1985, Cracking catalyst deactivation models, Ind. Eng. Chem. Prod. Res. Dev., 24, 10-15.

John T.M. and Wojciechowski B. W., 1975, On identifying the primary and secondary products of the catalytic cracking of neutral distillates, J. Catal., 37, 240-250.

Lopes J.M., Lemos F., Ramoa Ribeiro F., Derouane G., Magnoux P., Guisnet M., 1994, Coke deposition on H-ZSM-20 and USY zeolites, Appl. Catal. A, 114, 161-172.

Magnoux P., 1987, Mode de formation du coke et de desactivation des zéolithes-influence de la structure poreuse et de l'acidité, Thèse de doctorat de l'université de Poitiers.

Nace D.M., Voltz S.E. and Weekman, 1971, Application of a kinetic model for catalytic cracking- Effects of charge stocks, Ind. Eng. Chem. Proc. Des. Dev., 10, p. 530-537.

Pitault I., Nevicato D., Forissier M. and Bernard J.R., 1994, Kinetic model based on a molecular description for catalytic cracking of vacuum gas oil, Chem. Eng. Sci., 49, 4249-4262.

Turlier P., Forissier M., Rivault P., Pitault I. and Bernard J.R., 1994, Catalyst fouling by coke from vacuum gas oil in FCC reactors, "Fluid Catalytic Cracking III: materials and processes", Am. Chem. Soc. Symp. Ser. 571, ed. by Ocelli M.L. and O'Connor P., Washington, DC, 98-109.

Upson LL., Dalin I. and Wichers W.R., 3rd Katalistiks FCC Symposium, May 1984, Amsterdam.

Upson L.L, C. L. Hemler and D. A. Lomas, 1993, Unit design and operational control: impact on product yields and product quality, Fluid Catalytic Cracking: Science and Technology, Studies in Surface Science and Catalysis, Vol. 76, ed. by J.S. Magee and M. M. Mitchell, Elsevier, Amsterdam, 385–440.

Venuto P. B. and E. T. Habib, 1979, in FCC with zeolite catalysts, ed. by H. Heinneman, Dekker, New York.

COMPOSITE MOLECULAR SIEVE COMPRISING MCM-41 WITH INTERPOROUS ZSM-5 STRUCTURES: A NOVEL POROUS MATERIAL IN FCC PREPARATION

K. R. Kloetstra,[a] H.W. Zandbergen[b], J.C. Jansen[a] and H. van Bekkum[a]

[a]Laboratory of Organic Chemistry and Catalysis, Delft University of Technology, Julianalaan 136, 2628 BL Delft, The Netherlands

[b]Laboratory for Materials Science, Delft University of Technology, Rotterdamseweg 137, 2628 AL Delft, The Netherlands

Abstract

The recrystallization of MCM-41 into ZSM-5 is described. X-ray diffraction on mild TPAOH treated MCM-41 shows the vanishing of the MCM-41 d-spacings and the appearance of a peak with a much larger d-spacing. Besides, ZSM-5 peaks appear in the X-ray diffraction pattern. TEM measurements reveal that the original MCM-41 framework is affected and hardly intact over the whole volume. The tetrahedrally coordinated aluminum in the composite MCM-41/ZSM-5 (denoted as CMZ) gives a line narrowing in the ^{27}Al MAS NMR spectra compared to the original MCM-41 indicating an increase of the symmetry around the framework aluminum species. In contrast to the parent MCM-41, CMZ is active in hexane cracking at 450 °C. Intergrowths of MCM-41 and ZSM-5 deserve study as possible additives to enhance the octane value of gasolines generated by fluid cracking catalysts.

INTRODUCTION

The combination of macropores from an amorphous matrix with micropores of a zeolite is used today in FCC preparation (1). Controlled growth of macro-/mesopores ending up progressively in zeolitic micropores will probably enhance the accessibility of the various catalytic sites and thus the catalytic performance (1). In the field of zeolite synthesis such "cascade-type" pore configurations are obtained by preparing mesopores interconnecting micropores. In microporous materials such pore configurations can be achieved in various ways. Firstly, by creating mesopores via dealumination of Y (2), mordenite (3) or mazzite (4). Secondly, by pillaring of MCM-22 (5). Finally, by overgrowth of MCM-41 on FAU (6). A reversed way in preparing these meso-/microporous materials could be by creating micropores in a mesoporous host material. The flexibility of MCM-41 (7,8) in tuning its unidimensional pores in the range of 15-100 Å and its Si/Al ratio between 10 and infinity, makes this mesoporous alumino-silicate an excellent starting material. By combining the experience in the crystallization of ZSM-5 (9) and in the synthesis of MCM-41 (6,10), we explored a new route in preparing composite materials with both mesopores and micropores. Here we report a study of tetrapropylammonium hydroxide treatment of alumino-silicate MCM-41 with formation of a composite molecular sieve containing both MCM-41 and ZSM-5 structures. Materials of this type could be of particular interest to refiners. In fact, additives containing an estimated ≤3 wt% of ZSM-5 have been in use worldwide to improve gasoline octane (11,12). Typically the zeolite-containing additive, consists of a matrix such as alumina or an aluminosilicate gel (12) in which ZSM-5 crystallites are uniformly dispersed.

The approach chosen to synthesize the present composite molecular sieve is based on optimization of the ZSM-5 stoichiometry in the MCM-41 phase. It is well-known that in as-synthesized ZSM-5 each tetrapropylammonium (TPA) is surrounded by 24 tetrahedrally coordinated silicon and/or aluminum atoms (13). Thus in MCM-41 with a Si/Al of 30 there are thirty T-atoms neighbouring one TPA-cation in TPA-MCM-41. This ZSM-5 precursor-like situation is expected to accelerate ZSM-5 crystallization. Eventually, to obtain a mesoporous support with randomly very small microporous ZSM-5 crystallites of a few unit cells it is necessary to control the ZSM-5 synthesis and suppress the complete consumption of the MCM-41 material by interrupting ZSM-5 crystallization.

Crystallization of ZSM-5 in a granulated silica gel support resulted in small crystals of ZSM-5 with a size correlating with the pore size of the support (14). In our case we might expect a similar situation that the growth of the ZSM-5 crystallites is limited by the size of the unidimensional MCM-41 support.

EXPERIMENTAL

Synthesis

MCM-41 was prepared by mixing 16.5 g TMA-SiO$_2$ solution (TMA/SiO$_2$ = 0.5; 10 wt% SiO$_2$; tetramethylammonium hydroxide (TMAOH) was purchased from Aldrich and the silica source was Cab-osil M5 purchased from Fluka) with 6.35 g of sodium silicate (Aldrich; 27 wt% SiO$_2$). Subsequently 45 g of water and 4.7 g of silica (Cab-osil M5) were added to the mixture. Under vigorous stirring 4.9 g of cetyltrimethylammonium bromide (ACROS) in 100 g of water was poured in the mixture. Finally, the gel was enriched with 0.37 g of sodium aluminate (Riedel de Haën; 54% Al$_2$O$_3$). The resulting MCM-41 gel was put in a 250 mL polypropylene bottle and kept for 3 days at 100 °C. The solid was filtered and thoroughly washed, dried at 90 °C under vacuum and then calcined at 540 °C for 10 h under air. According to elemental analysis (ICP, AAS) the Si/Al ratio was 30.

The calcined MCM-41 (1.0 g) was allowed to stabilize with TPAOH (CFZ; 40 wt% aqueous solution) and water under stirring at room temperature for 2 h in the following molar ratios: 60 SiO_2 : Al_2O_3 : x TPA_2O : 1500 H_2O (x = 1 to 4). The mixture was put in a 15 ml teflon lined autoclave and allowed to react in a conventional oven at 170 °C for appropriate times. The resulting materials were cooled quickly to room temperature, collected by filtration, washed with distilled water and dried at 90 °C under vacuum. Eventually, the materials were calcined at 500 °C for 5h under air.

Characterization

The calcined samples were characterized by powder X-ray diffraction on a Philips PW 1840 diffractometer using monochromated CuKα radiation. Patterns were recorded from 1° to 40° (2Θ) with a resolution of 0.02° and a count time of 1 s at each point.

Solid-state ^{27}Al MAS NMR spectra were recorded at room temperature on a Varian VXR-400S spectrometer, equipped with a Doty Scientific 5 mm Solids MAS Probe. A resonance frequency of 104.21 MHz, a recycle delay of 0.1 s, short 3 μs pulses (45° pulses of 3 μs), a spectral width of 50 kHz and a spinning rate of 6.2 kHz were applied. The lines were referenced to $Al(NO_3)_3$ (0 ppm). The materials measured were hydrated by equilibrating them with air.

*FT*ir spectroscopy was performed on an IFS 66 Bruker Spectrometer using the KBr pellet technique.

Multipoint BET surface areas and pore volumes were calculated from N_2 adsorption/desorption isotherms at -196 °C using a Quantachrome Autosorb 6 instrument. The samples were outgassed for 16 h in vacuum at 350 °C prior to use. Argon adsorption measurements in the micropore as well as in the mesopore region were conducted on a Micromeritics ASAP 2000 apparatus reconstructed with very low-pressure equipment. The method of Horváth and Kawazoe (15) was used to determine the pore size distribution.

Transmission electron microscopy was performed with a Philips CM 30 ST electron microscope with a field emission gun operated at 300 kV and equipped with an energy dispersive X-ray (EDX) elemental analysis system. The ground samples were suspended in ethanol. A copper grid coated with a microgrid carbon polymer was loaded with a few droplets of this suspension.

The morphology of the crystalline aggregates was analyzed by using a Philips XL20 scanning electron microscope. The studied samples were first coated with a gold evaporated film.

Activity test

Samples of the parent MCM-41 and the TPAOH treated MCM-41 were tested in a hexane cracking test. The ammonia exchanged materials were carefully pelletized, crushed and sieved; the fraction with a diameter between 0.7 and 1.0 mm was collected. Cracking was performed at 450 °C and atmospheric pressure with 0.5 g of material stored under ambient, downflow, in a borosilicate glass tube (i.d. 7 mm) heated by a fluidized bed oven. The catalysts were pretreated under a nitrogen stream at 450 °C to accomplish the calcination of the NH_4-form to the H-form. The feed gas (400 µl N_2/s) contained 24.4 vol % hexane. Quantitative analysis was performed by on-line GC (cp sil5 column).

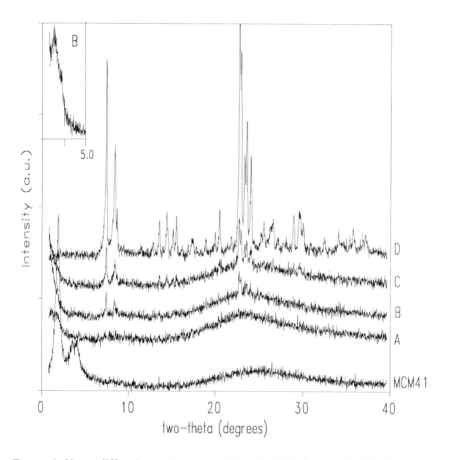

Figure 1. X-ray diffraction patterns of calcined TPAOH-treated MCM-41 materials (170 °C; x = TPA:Al = 4) with different recrystallization times: A = 90 min, B = 135 min, C = 180 min and D = 1080 min and parent MCM-41 (lower pattern).

RESULTS AND DISCUSSION

Preparation and characterization

Studying the TPAOH treatment of MCM-41 by X-ray powder diffraction some interesting aspects are observed. Figure 1 shows the X-ray diffractograms of the parent MCM-41 and the TPAOH-treated MCM-41 samples (calcined at 500 °C) at reaction times varying from 90 min to 1080 min (samples A to D; x = 4). It can be observed that the high order reflections of the parent MCM-41 (d_{110} = 22 Å and d_{200} = 19 Å) have disappeared. The intensity of the d_{100}-spacing (38 Å) of MCM-41 is moderately present in samples A and B. Surprisingly, a new peak with a 63 Å d-spacing can be observed in sample A. Sample B shows also a peak in the very low angle region related to a d-spacing of 67 Å. ZSM-5 peaks appear when the reaction time is around 135 min. Further increase of the reaction time leads to complete formation of ZSM-5 (sample D). In other words MCM-41 is recrystallized into ZSM-5. The strong decrease of the diffraction peaks of the original MCM-41 suggests that the features of MCM-41 have disappeared and the framework is destroyed and becomes amorphous. Sample A shows that the original phase is partly converted into another phase as indicated by the presence of a new peak with a spacing of 63 Å in the X-ray powder diffractogram and by TEM (*vida infra*), showing non-crystalline areas (indicated by arrows in Figure 3). Further quantification of these observations are too speculative however, this phase transformation is tentatively assigned to the development of ZSM-5 precursors. The gradual increase of the ZSM-5 X-ray diffraction peaks of the samples B and C indicates growing of ZSM-5 aggregates. Actually, the crystallization of ZSM-5 is interrupted in these samples. The TPA-containing sample D (uncalcined) shows an X-ray diffraction pattern (not shown) with a low intensity of the 2θ peaks at 7.8° and 8.7° compared to the 23.1° 2θ peak. This pattern differs from that shown in Figure 1 and that of ZSM-5 prepared in a homogeneous solution (16).

*FT*i.r. spectroscopy reveals that going from sample A to D the vibration at 550 cm^{-1} is gradually growing, indicating development of ZSM-5 skeleton structures (17). Though sample A shows a weak skeleton bending vibration at 550 cm^{-1} it does not possess an X-ray diffraction pattern. This suggests that this material contains very small ZSM-5 structures of a size at most of one unit cell. Differences between X-ray diffraction and i.r. crystallinity for ZSM-5 on an amorphous matrix have already been reported by other workers (16). It was already mentioned that the disappearance of the X-ray diffraction pattern of the original MCM-41 suggests at first sight a very fast consumption of MCM-41. However, when applying x = 1 and 90 min reaction time the consumption of MCM-41 could be controlled much better and the disappearance of the d-spacings is retarded.

Transmission electron microscopy (TEM) examination of sample A revealed the presence of 3 types of MCM-41 like particles (Figure 3). The first type, the primary phase, showed only barely the d_{001} spacing of MCM-41 in some areas, whereas this spacing was absent in other areas of the crystal. The second type showed the d_{001}

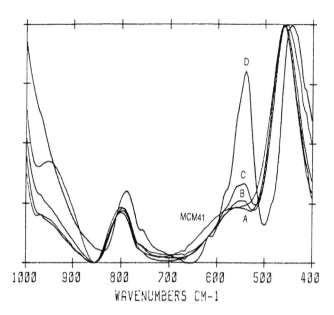

Figure 2. FTir absorption spectra of the samples A to D and MCM-41. The increase in intensity of the 550 cm^{-1} band can be observed.

spacings of MCM-41 rather dominantly, but with some patches in which these spacings were
(almost) absent. Figure 3 illustrates the MCM-41 framework which is severely affected by the hydrothermal treatment. Also voids and/or amorphous material (a few are indicated by arrows) can be observed. The third MCM-41 type material observed is crystalline with continuous d_{001} spacings. The two latter types were minor phases. Only the first type of MCM-41 particles were observed in samples B and C. These observations suggest that, within one batch several stages of the transformation of MCM-41 into another phase occur. The TEM experiments also show that, patches in which MCM-41 d_{001}-spacings were absent, do occur in a random manner in the whole volume representing the MCM-41 crystal.

The importance of the interporous framework-TPA (or aluminum) distance in MCM-41 is reflected by the difficulty in converting MCM-41 with a Si/Al = 20 into ZSM-5. In contrast to the MCM-41 sample described above (Si/Al=30) only a very small amount of X-ray crystalline ZSM-5 could be observed after TPAOH loading and reaction overnight at 170 °C (x = 4). TEM analysis of these materials reveals just slightly affected MCM-41 particles. Generally, a thermally treated aluminosilicate gel (sodium free) with a comparable Si/Al ratio is crystallized with difficulty towards ZSM-5 and leads to ZSM-5 of a poor crystallinity (18).

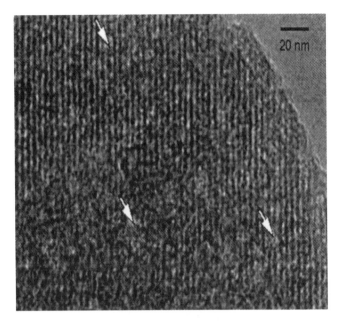

Figure 3. TEM image of sample A. Voids or amorphous areas are indicated by arrows.

Figure 4. SEM micrograph of sample D (enlargement 2730x).

As mentioned earlier, complete consumption of MCM-41 leads to ZSM-5. It appeared that the obtained ZSM-5 possessed an X-ray diffraction pattern which indicates the presence of aggregates with different crystal shape. Scanning electron microscopy (SEM) on sample D reveals that the morphology of the crystal aggregates is not uniform. The material consists, in some cases very clearly, of intergrown disks with sizes in the range of 3 to 7 μm (Figure 4). These crystal shapes resemble those of reported silica-supported small crystals of ZSM-5 (14). It can be concluded that the TPA-MCM-41 system promotes ZSM-5 crystal growth in a preferential direction (9).

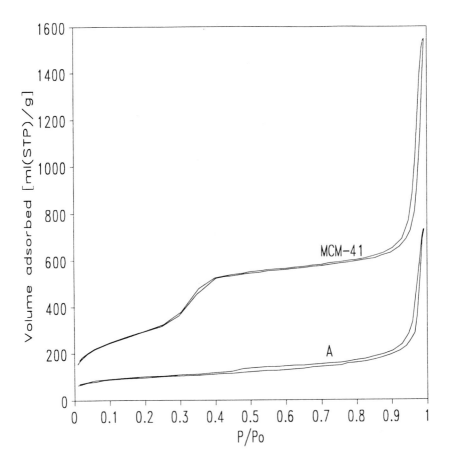

Figure 5. Nitrogen adsorption/desorption isotherms of the parent MCM-41 material and sample A.

Nitrogen physisorption on sample A differs substantially from that found for MCM-41 (Figure 5). MCM-41 shows the typical Type IV isotherm with the presence of hysteresis (7,8). The effect of the thermal TPAOH-treatment for 90 min on the pore aperture of MCM-41 is illustrated by the nitrogen isotherm of sample A. The sharp inflection at p/p_o = 0.3 - 0.4, characteristic of capillary condensation within uniform cylindrical pores, has disappeared (7). The desorption isotherm inflects to the adsorption isotherm at p/p_o = 0.45. This isotherm can be classified as a Type I isotherm with a hysteresis loop of the Type H4 of the IUPAC classification (19). The combination of this type of isotherm and the hysteresis loop is indicative of microporosity with large holes or slit-like pores (19,20). TEM data does not allow distinction between these

Table 1. Physisorption data of the parent MCM-41 material and sample A.

Sample C-value BET equation	BET area (m²/g)	Cumulative pore volume (cm³/g)
MCM-41 70.6	1079	0.802
A 451.5	335	0.171

two possibilities. The BET area of sample A has decreased by a factor 3 compared to MCM-41 (Table 1). The high *C*-value for the BET equation for sample A is normally associated with micropore filling (18). Micropore filling into cavities of molecular size can not be excluded, neither can secondary micropore filling which involves quasi-multilayer formation (20).

Horváth-Kawazoe (15) analysis on the parent MCM-41 and sample A shows again the drastic influence of the TPAOH treatment on the pore size distribution of MCM-41 (Figure 6). The Horváth-Kawazoe plot of MCM-41 shows two peaks. The peak at 10 Å results from monolayer adsorption and the pore filling peak is situated at 40 Å (21,22). It is observed that sample A possesses micropores, with a maximum at 5.5 Å which is in harmony with ZSM-5 formation. The distribution is spread out gradually to the mesopore region indicating the presence of a broad distribution of mesopores.

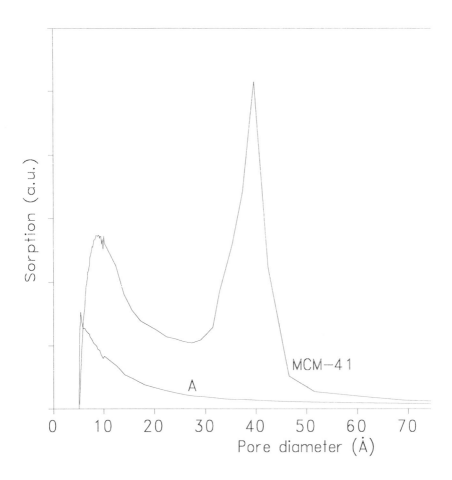

Figure 6. Horváth-Kawazoe plots of the parent MCM-41 and sample A.

Figure 7 shows the ^{27}Al MAS NMR spectra of the parent MCM-41 and of samples A to D. Deconvolution of the ^{27}Al NMR spectrum of MCM-41, assuming Lorentzian profiles for each component, reveals a splitting of the octahedral peak into two peaks at 30 ppm and -2 ppm. The former peak is tentatively assigned to extra-framework aluminum (23) and the latter peak to octahedral aluminum. Compared to the original MCM-41 a large reduction in extraframework aluminum is observed in samples A, B and C. Although effects of the aluminum quadrupole moment and the quadrupole coupling constant (24,25), cannot

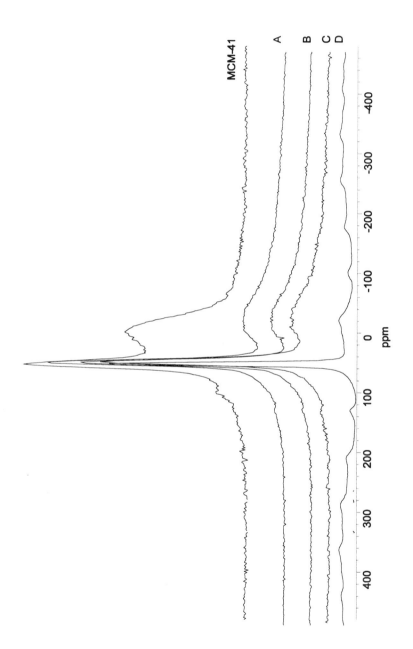

Figure 7. ^{27}Al MAS NMR spectra of the parent MCM–41 and of the samples A to D.

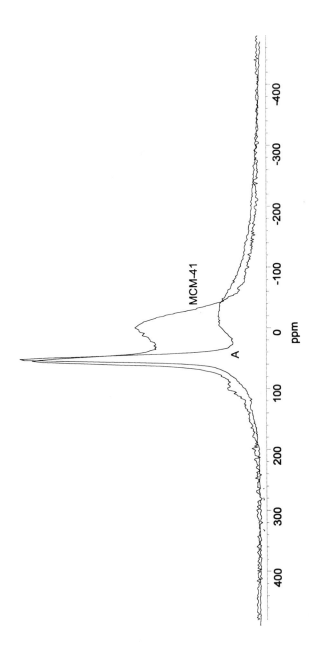

Figure 8. ^{27}Al MAS NMR spectra of the parent MCM-41 and sample A.

be excluded, the decrease of the broad peak is attributed to extraframework aluminum (26) conversion during ZSM-5 formation. Sample D shows only one peak at 53 ppm which can be assigned to tetrahedral aluminum (24). Line-narrowing of the tetrahedral aluminum peak in the samples A to D is noticed (Figure 8). This indicates that the local homogeneity of the framework AlO_4 species in the TPAOH treated samples has increased indicating a smaller T-O-T angle distribution (T represents a SiO_4 or AlO_4^- tetrahedron) (18,23,24) due to formation or arrangement of double five-membered rings (27).

Activity testing

Hexane cracking is an often used technique to investigate the strength of the Brønsted acid sites of a catalyst (28,29) or the crystallinity of ZSM-5 (30). Generally, only strong Brønsted acid sites are capable of cracking hexane at mild conditions (350 °C - 450 °C). Table 2 illustrates the performance of the studied samples in cracking hexane at 450 °C in a continuous flow reactor. The samples A, B and C appeared to be active in cracking hexane, though the conversions are rather low compared to ZSM-5 (sample D). It is noteworthy, that a commercial aluminosilicate catalyst containing 20

Table 2. Hexane cracking activity of tested samples.

Sample BET surface area (m²/g)		Conversion (μmol hexane/g.min)[a]	
D		354.5	-
C		3.9	-
B		1.4	-
A	335	2.4	
MCM-41	1079	0	
HA-HPV[b]	433	0	

a) Conditions: flow = 400 μL N_2/s, 24.4 vol % hexane and T = 450 °C. b) Aluminosilicate with 25 wt% alumina and a pore volume of 0.87 ml/g, provided from Akzo-Nobel Chemicals BV.

wt% of alumina tested under the same conditions showed no hexane cracking activity. Neither did the parent MCM-41 material which contains relatively weak acid sites (25). The measured activity of sample A compared to sample B and C which both contain small ZSM-5 aggregates suggests that this cracking activity of sample A is caused by the small amount (about 10%) of affected MCM-41 material (Figure 3). The affected framework of MCM-41 contains strong acid sites presumably from very small interporous ZSM-5-type aggregates with a size less than one unit cell. The activity of sample A might be improved substantially if the other 90% of the material would also be accessible for catalysis. It appeared that all the samples showed rapid deactivation by coke formation. The observed activities are consistent with the reported activity for ZSM-5 particles with a size of a few unit cells on an amorphous matrix (18). The low activity of samples B and C, possessing X-ray crystalline ZSM-5, indicates a poor accessibility of the ZSM-5 aggregates which are surrounded by unconverted amorphous material. A comparable low accessibility of silica-supported ZSM-5 crystals in cracking reactions is found by Landau *et al.* (14).

Sample A shows most of the features of CMZ, a composite MCM-41/ZSM-5 molecular sieve, but the amount of amorphous material has to be decreased by controlling the consumption of MCM-41. Optimizing the experimental conditions might then lead to a material which possesses the real characteristics aimned at.

Our approach to synthesize CMZ is actually the opposite of the procedure of the reported amorphous mesoporous silica-alumina preparation of Bellussi *et al.* (18). These authors interrupted the ZSM-5 synthesis and obtained ZSM-5 precursors which possessed a mesoporous size distribution. However, these systems were achieved within a small crystallization-time window and showed never an X-ray diffraction crystallinity of ZSM-5. Thus, the system of Bellussi *et al.* might be more difficult to control.

CONCLUSIONS

An explorative study on mild tetrapropylammonium hydroxide treatment of MCM-41 shows that a porous aluminosilicate MCM-41 material with ZSM-5 structures is obtained. The phase transformation occurs randomly, via several stages, in the whole MCM-41 material. The recrystallization of MCM-41 into ZSM-5 is stimulated by applying the optimized stoichiometry of ZSM-5 in TPA-MCM-41, leading to short crystallization times. The CMZ materials show a higher activity in cracking hexane than the original MCM-41 indicating generation of strong Brønsted acid sites. Nevertheless, the synthesis of CMZ has to be improved in order to obtain a molecular sieve with a large amount of accessible active sites. Materials of this type should be studied as possible octane enhancing additives in FCC preparation.

ACKNOWLEDGMENT
We thank P.J. Kunkeler for assistance with the cracking experiments, J.C. Groen for the physisorption data and A. Sinnema for carrying out the NMR measurements. This research is supported by NIOK, the Dutch School of Catalysis.

LITERATURE CITED
(1) O'Connor, P. and Humphries, A.P. *Symposium on Advances in Fluid Catalytic Cracking, ACS National Meeting*, 598 (1993).
(2) Patzelová, V and Jaeger, N.I. *Zeolites*, 7, 240 (1987).
(3) Lee, G.S., Garcés, J.M., Meima, G.R., Van der Aalst, M.J.M. *Eur. Pat.*, 0.433.932 (1990).
(4) Buckermann, W.A., Huong, C.B., Fajula, F., Gueguen, C. *Zeolites*, 13, 448 (1993).
(5) Roth, W.J., Kresge, C.T., Vartuli, J.C., Leonowicz, M.E., Fung, A.S. and McCullen, S.B. *Stud. Surf. Sci. Catal.*, 94, 301 (1995).
(6) Kloetstra, K.R., Zandbergen, H.W., Jansen, J.C. and Van Bekkum, H. *Microporous Mater.*, 6, 287, (1996).
(7) Kresge, C.T., Leonowicz, M.E., Roth, W.J., Vartuli, J.C. and Beck, J.S. *Nature*, 359, 710 (1992)
(8) Beck, J.S., Vartuli, J.C., Roth, W.J., Leonowicz, M.E., Kresge, C.T., Schmitt, K.D., Chu, C.T-W., Olson, D.H., Sheppard, E.W., McCullen, S.B., Higgins J.B. and Schlenker, J.L. *J. Am. Chem. Soc.*, 114, 10834 (1992).
(9) Jansen, J.C., Kashchiev, D. and Erdem-Senatalar, A. *Stud. Surf. Sci. Cat.*, 85, 215 (1994).
(10) Kloetstra, K.R., Zandbergen, H.W. and Van Bekkum, H. *Catal. Lett.*, 33, 157 (1995).
(11) Schipper, P.H., Dwyer, F.G., Sparrel, P.T., Mizrahi, S. and Herbst, J.A. in *Fluid Catalytic Cracking: Role in Modern Refining.*, M.L. Occelli Ed., ACS Symposium Series No. 375, p. 64 (1988).
(12) Miller, S.J. and Hsieh, C.R. in *Fluid Catalytic Cracking II: Concepts in Catalyst Design.*, M.L. Occelli Ed., ACS Symposium Series No. 452, p. 96 (1991).
(13) Nagy, J.B., Gabelica, Z. and Derouane, E.G. *Zeolites*, 3, 43 (1983).
(14) Landau, M.V. and Herskowitz, M. *Stud. Surf. Sci. Catal.*, 94, 357 (1995).
(15) Horváth, G. and Kawazoe, K. *J. Chem. Eng. Jpn.*, 16, 470 (1983).
(16) Kokotailo, G.T., Lawton, S.L., Olson, D.H. and Meier, W.H. *Nature*, 272, 437 (1978)
(17) Jacobs, P.A., Derouane, E.G. and Weitkamp, J. *J. Chem. Soc., Chem. Commun.*, 591 (1981).
(18) Bellussi, G., Perego, C., Carati, A., Peratello, S., Previde Massara, E., Perego, G. *Stud. Surf. Sci. Catal.*, 84, 85 (1994).
(19) Sing, K.S.W., Everett, D.H., Haul, R.A.W., Moscou, L., Pierotti, R.A., Rouquérol, J. and Siemieniewska, T. *Pure & Appl. Chem.*, 57, 603 (1985).

(20)   Sing, K.S.W. *Colloid Surf.*, 38, 113 (1989).

(21)   Vartuli, J.C., Kresge, C.T., Leonowicz, M.E., Chu, A.S., McCullen, S.B., Johnson, I.D., Sheppard, E.W., *Chem. Mater.*, 6, 2070 (1994).

(22)   Vartuli, J.C., Schmitt, K.D., Kresge, C.T., Roth, W.J., Leonowicz, M.E., McCullen, S.B., Hellring, S.D., Beck, J.S., Schlenker, J.L., Olson, D.H., Sheppard, E.W., *Chem. Mater.*, 6, 2317 (1994).

(23)   Busio, M., Jänchen, J. and Van Hooff, J.H.C. *Microporous Mater.* 5, 211 (1995).

(24)   Oldfield, E., Haase, J., Schmitt, K.D. and Schramm, S.E. *Zeolites*, 14, 101 (1994).

(25)   Hunger, M. and Horvath, T. *Ber. Bunsenges. Phys. Chem.*, 99, 1316 (1995).

(26)   Corma, A., Fornés, V., Navarro, M.T. and Pérez-Pariente, J. *J. Catal.*, 148, 569 (1994).

(27)   Scholle, K.F.M.G.J., Veeman, W.S., Frenken, P. and Velden, G.P.M., *Appl. Catal.*, 17, 233 (1985).

(28)   Olson, D.H., Haag, W.O. and Lago, R.M. *J. Catal.*, 61, 390 (1980).

(29)   Haag, W.O., Lago, R.M. and Weisz, P.B. *Nature*, 309, 589 (1984).

(30)   Hardenberg, T.A.J., Mertens, L., Mesman, P., Muller, H.C. and Nicolaides, C.P. *Zeolites*, 12, 685 (1992).

HYDROGEN-TRANSFER ACTIVITY OF MCM-41 MESOPOROUS MATERIALS IN 1-HEXENE AND CYCLOHEXENE ISOMERIZATION REACTIONS

J. M. Domínguez, F. Hernández, E. Terrés, A. Toledo, J. Navarrete and M. L. Occelli[*]
Instituto Mexicano del Petróleo, Eje Central Lázaro Cárdenas 152, STI, Gcia. Catálisis y Materiales, 07730 México D.F.
*Georgia Tech. Research Institute, Atlanta GA 30332-0800, U.S.A.

Mesostructured aluminosilicates of the MCM-41 type have been synthesized using different sources of silica and surfactants. Following the thermal decomposition of the organic template, the crystals were found to contain extraframework Al, had surface area in the 700-1100 m^2/g range and pore diameter in the 2.4-4.5 nm range. Pore size and pore size distribution have been investigated also by HREM.

FTIR experiments with chemisorbed pyridine indicate that these MCM-41 crystals, in the presence of Na and high levels of Al(VI) species, contain only Lewis type acidity. Acid site strength is weaker then the one observed in pillared clays and resemble the one in amorphous aluminosilicates.

Isomerization reactions have been used to investigate the hydrogen transfer index (HTI) of these mesoporous materials. Results from 1-hexene conversion at 400°C give methylpentane/methylpentenes ratio that are similar to the one obtained when using HZSM-5 crystals but are 1/10 smaller then the one from a HY zeolite of the type used in commercial FCC. Similarly, when converting cyclohexene with these MCM-41, HTI values are significantly lower then the one obtained with HZSM-5 and HY zeolites.

Hydrothermal stability as well as coke selectivity will have to be drastically improved before these materials can be considered in FCC applications.

175

I. INTRODUCTION

The lead phaseout, and the worldwide adoption of ever stringent automotive emission standards, have focused the petroleum engineer attention on oxygenates generation and availability. The demand for reformulated gasoline with reduced aromatics will require the use of additives such as MTBE (methyl-ter-buthyl ether), TAME (ter-amyl-methyl ether), ETBE (ethyl-ter-buthyl ether), and DIPE (di-isopropyl ether). These additives can be produced from the etherification of certain iso-olefins $\left(i\text{-}C_3^=, i\text{-}C_4^=, i\text{-}C_5^=\right)$ with light alcohols (methanol or ethanol). Natural gas and crude oil are the traditional raw materials for the alcohols and olefins needed for the oxygenates used today as octane enhancers.

Since 1991, in the USA, major cities with high CO levels have been required to use gasoline containing 2.7 wt% oxygen during the winter season. At the present, MTBE remains the preferred mean to meet mandates of this type. The large scale production of MTBE is based on the reaction of methanol with isobutylene. Currently, isobutylene is produced by several well established processes including butane dehydrogenation. However, new sources of isobutylene will be needed to satisfy rising demands for MTBE usage in motor fluids. MTBE production capacity is projected to increase to 16,141,000 ton in the year 2000.

Because methanol does not react with saturates, n-olefins or di-olefins, and react only with tertiary olefins, any available C4-stream from a FCCU, containing isobutylene can be used as a feed source for MTBE generation. Thus refiners interest in finding new FCC compositions that can improve olefin yields from a FCCU, remains high (1). New catalysts (such as Engelhard IsoPlus 2000) and processes (such as Mobil and BP Isofin process and UOP Ethermax process) have already appeared .

Traditional zeolite-containing FCC promote hydrogen transfer reactions during gas oil cracking. As a result paraffin yields have remained high (2). New FCC composition with low hydrogen transfer index (HTI) are needed to promote olefin formation over paraffins generation from FCCU. Recently (3,4) it has been reported that the HTI during propylene conversion decreases with increasing the zeolite Si/Al ratio. Process variables such as T (i.e T>400C) and space velocity (i.e. WHSV>10H^{-1}) can lower the catalyst HTI probably because of a mechanism of rapid desorption of the olefinic intermediates (3).

MCM-41 materials represent a new family of aluminosilicates with unique pore diameters in the 2 nm to 10 nm range and stability in air to 800°C (5). They can contain both Bronsted (B) and Lewis (L) acid sites (6,7). Acid site strength is comparable to that of amorphous aluminosilicates, but it is weaker then the one measured in HY type zeolites (8). Surface acidity may be enhanced by sulfonation or fluorination (9) or quenched by the presence of charge compensating Na ions.

It the purpose of this paper to report the hydrogen transfer properties of MCM-41 type materials during 1-hexene conversion at 400°C and during cyclohexene conversion at 250°C-350°C. Results will be compared with those from zeolites of the type used in FCC preparation. The HTI will be defined as the methylpentanes/methylpentenes and methylcyclopentanes/ methylcyclopentenes ratio, respectively.

II.   EXPERIMENTAL

(A)   Synthesis of MCM-41

Two methods were followed in the preparation of single phase MCM-41 mesoporous materials:

S1: Sodium aluminate and sodium silicate (Aldrich, 14% NaOH and 27% SiO_2) were added sequentially to a very diluted sulfuric acid solution (with stirring). Then, a mixture of 5.6g. of Cetyltrimethylammonium Bromide (CTMA-Br from Aldrich, 99%) with demineralized water (23.3 ml) was added to form a hydrogel having the following molar oxide composition:

$$Al_2O_3 : 8.64\ SiO_2 : 4.74\ (CTMA)_2 : 631\ H_2O$$

This mixture was heated at 140°C for 48 hours, in an autoclave. After filtering and drying (25°C, 12 hours), the solids were first calcined in nitrogen atmosphere for 4 hrs, then for 6 hrs in oxygen at 540°C (10).

S2: A volume of CTMA-Br (Aldrich, 99%), equivalent to 10 g in 30 g water, were exchanged by means of an anionic resin (Amberlyst 400); then 0.4 g of alumina (Catapal, Vista Co.) was added and the mixture stirred. After 10 min, 20 g of triethylammonium (TEA Aldrich, 99%) silicate together with 5 g of silica (Merck) were added to the mixture to form the following hydrogel:

$$Al_2O_3 : 10.93\ SiO_2 : 6.99\ (CTMA)_2O : 425\ H_2O$$

The hydrogel was then heated in a autoclave at 150°C for 48 hours (11). After filtering and drying (60°C, 12 hours) the solids were calcined using the conditions described above.

Reference HY and USY crystals were obtained from Davison. The HZSM-5 sample was instead obtained from Conteka

(B) Characterization

X-ray diffraction studies were carried out with a Siemens D500 diffractometer fitted with a graphite monochromator, λ (CuK_a) = 1.54 Å. The angular scale was expanded to have a better resolution of the low angle region. High resolution microscopy (HREM) was performed with a JEOL-2010 microscope, fitted with Si/Li Noran X-ray detector.

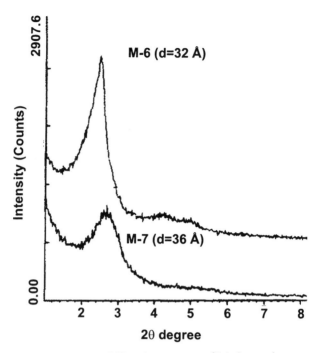

Figure 1. X-ray diffraction pattern of M-6 sample.

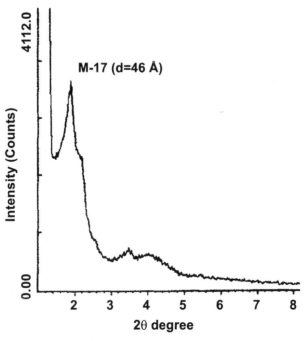

Figure 2. X-ray diffraction pattern of M-17 sample. M-17 crystals have been prepared from twice OH-exchanged CTMA-Br.

Surface area, pore size distribution and total pore volume were determined with an ASAP-2000 Porosimeter from Micromeritics. Average pore diameter (APD) data was obtained using the BJH method (17). Infrared (IR) spectra were recorded using a Nicolet 7000 spectrometer and KBr pressed discs. Pyridine was adsorbed at room temperature and the cell heated up to 400°C, in order to determine the relative acid strength of the acidic sites present. The amount of pyridine adsorbed either on Bronsted or Lewis sites was determined by the Lambert-Beer law (11) which establish a direct relationship between band intensities areas and the adsorbate (pyridine) concentration. The integrated molar extinction coefficient used was 3.03 cm/μmol for B-sites and 3.26 cm/μmol for L-sites (18,19). In addition, NH_3-TPD experiments were used to determine the total acidity of the solids.

(C) Reaction conditions

The HTI of MCM-41 materials was determined by means of the 1-hexene and cyclohexene conversion, respectively, at 1 atm. in a continuous flow system using temperatures in the 250-400°C range. A pyrex U-type differential reactor fitted with a porous glass support was used. Very small amounts of catalyst, i.e. M_{cat}=0.1g, were used in order to mininize possible diffusional effects. The analysis was performed with a Varian Star 3700 CX GC, fitted with a FID and a PONA capillary column with dimensions 50 m x 0.2 mm. x 5 m (film thickness). The feed consisted of either 1-hexene (6 x 10^{-3} gmol/min, Vo=180 ml/min, T= -7°C) or cyclohexene (1.8 x 10^{-4} gmol/min, Vo=30 ml/min, T= 25°C) and helium was the carrier gas. The main reaction products were the isomers, i.e. methylpentenes, methylpentanes, 2-hexenes, n-hexane and light hydrocarbons in the former case and methylcyclopentane, methylcyclopentene, in the latter case. The Hydrogen Transfer Index (HTI) was defined as the ratio methylpentanes/methylpentenes or methylcyclopentane/methyl-cyclopentene, respectively.

III. RESULTS AND DISCUSSION

Table 1 summarizes the gel OH/Si ratios and the properties of the MCM-41 crystals obtained. Samples M-6 and M-7 have been prepared by the same synthesis procedure (S1) but using different surfactants. In fact, M-6 was prepared using dodecyltrymethylammonium (DTMA) ions and M-7 (as well as M-16 and M-17 samples) using cetyeltrymethyl ammonium (CTMA) cations; X-ray diffractograms are shown in Figures 1 and 2. In this case, by changing the surfactant size, the mean pore diameter of the crystals increased to 3.27 nm from 2.47 nm, Table 1,

Table 1 Synthesis parameters and crystal properties of some MCM-41 type materials.

Sample	Organic Template	OH/Si (Gel)	d(100) (Å)	APD (Å)	Δ (Å)	SA (m²/g)
M-6	DTMA	0.28	33.0	24.7	13.40	1033
M-7	CTMA	0.28	36.0	32.7	8.87	1152
M-16	CTMA	0.84	48.5	41.0	15.00	724
M-17	CTMA*	0.52	46.0	44.1	9.02	1012

* Twice OH exchanged

Sample M-16 was also prepared with the S1 method. Data in Table 1 indicate that by increasing the gel OH/Si ratio, it is possible to increase the size of the mesopores.

Sample M-17 was instead prepared using the S2 procedure in order to avoid inclusion of Na in the MCM-41 crystals. In preparing M-17, the CTMA-Br solution was twice exchanged over anionic resins (Amberlyst-400) to increase the gel OH/Si ratio. In Table 2, it can be seen that MCM-41 type crystals prepared by the S1 method contain Na. Furthermore, Na/Al>1.0 indicating that in addition to charge compensating Na ions, the crystals contain Na impurities resulting from reactants occlusion in the interparticle voids formed by crystallites agglomeration.

^{27}Al NMR data in Table 2 indicate that all samples contain extraframework Al and that Al(VI) levels increase with the gel OH/Si ratio. Furthermore, the spectrum of sample M-16, in addition to a resonance near 55 ppm and 5 ppm, contains a weak peak near 21 ppm that could be attributed to highly distorted Al atoms in a tetrahedral environment (20), or to Al(V) species (21). Because of their low Al(IV)/Al(VI) ratios, sample M-16 and M-17 could be considered to be Si-MCM-41 crystals containing some extraframework Al-species in their mesopores.

Typical morphology and aggregation state of the MCM-41 grains, is shown by the SEM images in Figure 3. Plates and plate aggregates irregular in size and shape are common in these materials. The high resolution electron microscopy (HREM) image in Figure 4 show the pore distribution in these mesoporous materials. The image for sample M-7, consist of an hexagonal arrangement of white spots representing mesopores about 4.0 nm in diameter. Some amorphous patches around the grains of the crystals are also present. It is beleaved that these MCM-41 materials are not strictly crystalline, but the periodic array of mesopores give them a crystalline character. In Figure 4, the black contrast areas between white spots, represent the condensed amorphous aluminosilicate phase that constitute the pore walls. The mean

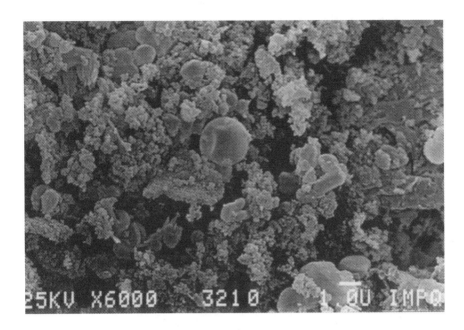

Figure 3. SEM micrographs showing the morphology and aggregation state of typical grains in MCM-41 type materials.

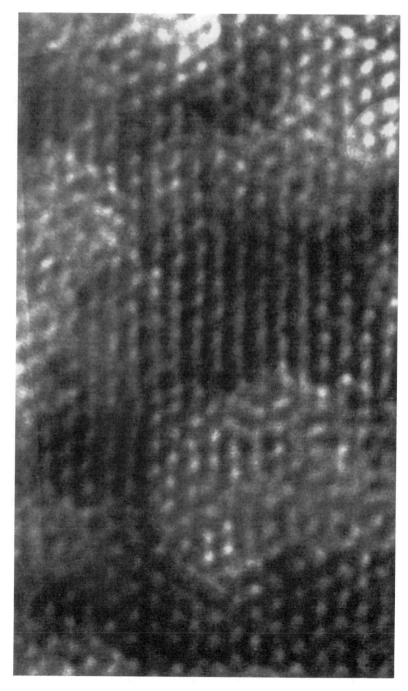

Figure 4. High resolution electron microscopy image of M-7 (1cm=36Å).

Table 2. Some acidity properties of the MCM-41 crystals described in Table 1.

SAMPLE	$\dfrac{Si}{Al}$	$\dfrac{Na}{Al}$	$\dfrac{Al(IV)}{Al(VI)}$	Acidity TPD-NH3 μmol/g	Acidity IR-PY μmol/g
M-6	6.2	3.7	2.68	74.8	71.0
M-7	6.7	3.7	1.90	185.5	107.0
M-16	22.0	1.2	0.05	45.8	57.0
M-17	11.4	0.0	0.23	178.6	186.0

wall thickness measured from this HREM image is about half the lattice periodicity measured by XRD.

By subtracting from the unit cell dimension $a_0 = 2d_{100}/\sqrt{3}$, the APD value, it is possible to estimate Δ, the thickness, of the pore wall. Results in Table 1 show the dependence of the wall thickness on synthesis conditions. As mentioned before, the mesopores APD increases with the size of the surfactant used but at the expense of the wall thickness and probably stability. On the other hand, in the presence of CTMA ions, variations in OH/Si ratios, (and Si/Al(IV)) afford both increases in APD and wall thickness; see samples M-16 and M-7 in Tables 1,2.

Surface acidity was investigated by recording IR spectra of chemisorbed pyridine and by performing NH3-TPD experiments, see Table 2 and Figure 5. Following the oxidative decomposition of the surfactant at 600°C, extraframework Al-species are formed (13,14). Other sites available to chemisorption of basic probe molecules results from Si-O-Al groups generated by Al insertion on the walls and into the walls of these mesostructured aluminosilicates.

In Table 2, sample M-6 and M-7, with similar Si/Al and Na/Al ratio, offer different chemisorption capacities probably because of differences in Al(VI)-levels and Al(IV) distribution (15). The decrease in NH3 uptake by sample M-16 has been attributed to the crystals large increase in the Si/Al ratio, Table 2. Finally, sample M-17 containing almost half the Al present in sample M-7 and with Al(IV)/Al(VI) = 0.23, can chemisorb as much NH3 and pyridine as sample M-7 probably because it is essentially free from Na impurities; see Table 2.

Since after calcination these materials have not been ion exchanged with NH$_4$-ions, it is believed that Bronsted sites will mostly be neutralized by charge compensating Na ions, and therefore acidity should be mainly of the Lewis type. Results in Figure 5 indeed show that in the presence of residual

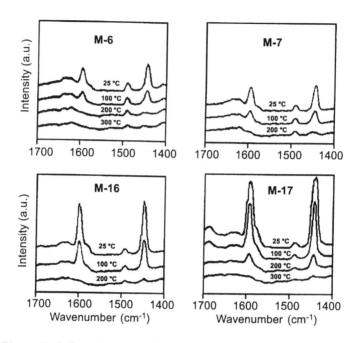

Figure 5. Infrared spectra of pyridine sorbed on MCM-41, see Table 1.

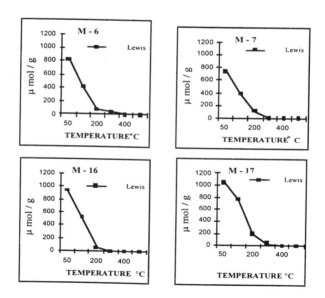

Figure 6. Variation of pyridine adsorption profile with temperature.

Na-ions, these MCM-41 contain only Lewis acid sites. The absence of pyridinium ions bands for sample M-17 in Figure 5 is somewhat surprising since in this sample Na/Al=0. It is believed that extraframework species such as AlO$^+$ prevent pyridine chemisorption on the crystals Bronstead sites. In all the samples examined, Figure 6 shows that thermodesorption of pyridine is essentially complete in the 200-300°C T range, indicating that acid site strengh is significantly lower then the one observed in pillared clays and zeolites (15), Figures 5,6.

Catalytic data from the isomerization reaction of 1-hexene at 400°C, is given in Table 3. These results indicate that, when compared at similar conversion, these MCM-41 material have HTI value similar to the one of HZSM5 (Si/Al=30) but it is about ten times smaller then the one for a HY type zeolite, Table 3. However HZSM5 and HY crystals are hydrothermally stable; these mesoporous materials are not.

Table 3. Hydrogen transfer properties of MCM-41 during 1-hexene conversion

CATALYST	TOTAL CONVERSION (% at 400 °C)	HYDROGEN TRANSFER INDEX
M-6	96.0	0.063
M-7	86.8	0.035
M-17	91.7	0.059
HZSM5	95.0	0.060
HY	94.7	0.764

Wc=100mg, vo=0.006 gmol/min
*Defined as the methyl-pentanes / methyl-pentenes ratio

Results from cyclohexene conversion are shown in Table 4 and Figures 7,8. In the 250-350°C range, conversion increases with T for all the materials tested. Furthermore, Figure 7 show that some catalyst deactivation occures with time and that activity losses are minimized at high conversion (reactor T=350°C). In fact, at 350°C, conversion seems to be weakly dependent on the crystals Si/Al ratio. In contrast, at 250°C and 300°C catalyst activity increases with Si/Al values; Figure 8. The lack of hydrogen transfer activity of these mesostructured aluminosilicates is further illustrated in Table 4. During cyclohexene conversion, HTI<<1.0 at all conversion levels and the HTI of a USHY (Si/Al=18.6) is much higher then the one reported for the MCM-41 materials under study; see Table 4.

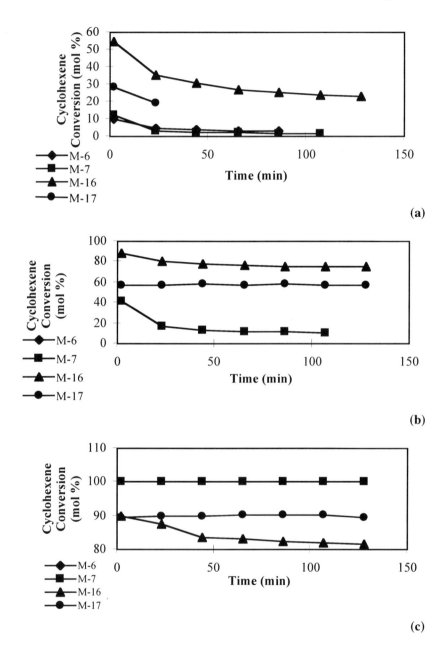

(a)

(b)

(c)

Figure 7. Variation of the cyclohexene conversion (mol %) with time on stream of three solids at (a) 250°C, (b) 300°C and (c) 350°C.

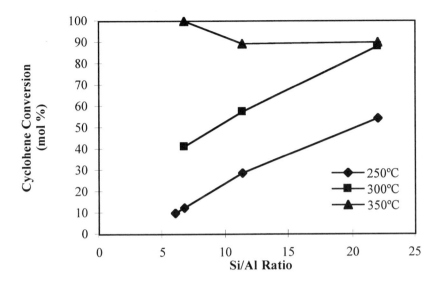

Figure 8. Variation of cyclohexene conversion with Si/Al ratio at different temperatures.

Table 4. Cyclohexene conversion as a function of temperature.

SAMPLE	TEMPERATURE (°C)	CONVERSION (%mol)	HTI*
M-6	250	28.6	0.12
M-7	250	12.2	0.05
M-7	300	41.4	0.11
M-7	350	100	0.55
M-16	250	54.6	0.09
M-16	300	88.4	0.08
M-16	350	89.9	0.06
M-17	250	28.6	0.0
M-17	300	57.6	0.0
M-17	350	89.5	0.01
USHY	250	95.3	70.8

Wc=100mg, Vo=0.006 gmol/min.
* Defined as the methylcyclopentanes / methylcyclopentene ratio

IV. SUMMARY AND CONCLUSIONS

The hydrothermal transformation of aluminosilicate hydrogels to mesoporous materials is controlled by the nature of the surfactant and by the gel Si/Al and OH/Al ratios. By replacing DTMA with CTMA ions, it is possible to increase the microstructure APD at the expense of the overall wall thickness. In the presence of CTMA ions, APD and wall thickness can be simultaneously increased by changing both Si/Al and OH/Si ratios in the starting hydrogel. Synthesis conditions can be changed to affect the levels of Al incorporated into the silicate framework. At high OH/Si ratio, large Al(VI)/Al(IV) values have been observed by ^{27}Al NMR.

Trends in NH3 and Pyridine chemisorption can be explained in term of Si/Al ratio, possible Al(VI)-species and Al-O-Si groups on the wall surface and inside the walls. FTIR of chemisorbed pyridine indicate that in these MCM-41, acidity is of the Lewis type probably because of the presence of charge compensating Na^{+} ions and high levels of extraframework Al(VI)-species. Pyridine thermodesorption is complete in the 200-300°C T-range indicating that acid site strength is weaker then in either zeolites or pillared clays.

The large (2 to 5 nm) APD of these moderately acidic mesostructures containing mainly Lewis acid sites, allow the transformation of 1-hexene and cyclohexene with very low HTI even at high conversion, probably because of short contact times and rapid desorption of olefinic intermediates. Thus , at least in principle, mesoporous materials of the type described in this report and elsewhere (5) could promote the retention of olefins during gas oil cracking.

It has been reported that aluminosilicate mesostructures, of the type described in this paper suffer from two major limitations (16). Although they have excellent thermal stability in air to 800°C, when exposed to 100% steam at 1 atm for five hours, they undergo a total structural collapse at T > 700°C (16).

At microactivity test conditions, it has been observed that prior to steam aging, MCM-41 crystals have gas oil cracking activity comparable to that of an equilibrium FCC (16); however, their coke/conversion ratio is 2 to 3 times higher(16). Thus hydrothermal stability as well as coke selectivity will have to be drastically improved before these materials can be considered in FCC applications.

V. REFERENCES

1.   J.B. Maclean, NPRA Annual Meeting, AM-92-45 (1992)
2.   W.C. Cheng , W.Suarez and G.W. Young, AlChE Symposium Series, No. 291, 88, 38 (1992)
3.   C.Mark Tsang, P.E. Dai, F.P. Mertens and R.H.Petty, Symp. HTR Hydrocarbon Process., Div.Petr.Chem.,Inc, 208th Natl. Meet. ACS, Washington D.C., Aug 21 (1994).
4.   J.Guan, Z. Yu, Y.Fu, C. Lee and X.Wang, Symp.HTR Hydrocarbon Process., Div. Petr. Chem., Inc.,208th Natl.Meet. ACS, Washington D.C., Aug 21 (1994)
5.   J.S. Beck, J.C.Vartuli, W.J.Roth, M.E. Leonowicz, C.T.Kresge, K.D. Schmitt, C.T.W. Chu, D.H., Olson, E.W. Sheppard, S.B. McCullen, J.B. Higgins and J.L. Schlenker, *J.Am.Chem.Soc.* (1992), 114, 10834-10843.
6.   A. Corma, V. Fornes, M. T. Navarro and J. Perez-Pariente, "Acidity and stability of MCM-41 aluminosilicates," in *Journal of Catalysis*, Vol. 148, pp. 569-574, 1994.
7.   R. B. Borade and A. Clearfield, "Preparation and characterization of acidic properties of MCM-41," *in Synthesis of porous materials: zeolites, clays and nanostructures*, edited by Mario L. Occelli and Henri Kessler, Marcel Dekker, Inc., New York, 1987.
8.   A. Corma, V.Fornés, M.T.Navarro and J. Perez-Pariente. *J. Catal.* , 148, 569-574 (1994).
9.   E. Terrés, J.M.Domínguez, M.A.Leyva, P. Salas and E. López, Symp. *Synthesis of Microporous Materials and Clays.*, In Press (1995), Marcel Decker Eds., Edited by M.L.Occelli.
10. R. Schmidt, D. Akporiaye, M.Stocker, O. Henrik Ellestad, in "Zeolites and related microporous materials: State of the Art, 1994."Studies in Surface Science and Catalysis, vol. 84, 61-67 (1994).
11. C.T. Kresge, M.E.Leonowicz, W.J.Roth, J.C. Vartuli and J.S. Beck, *Nature*, vol. 359, 710-712 (1992).
12. T. R. Hughs and H. M. White, *Journal of Physical Chemistry*. 71, 7 (1967)
13. B. Borade Ramesh and A. Clearfield, "Synthesis of aluminum-rich MCM-41," in *Catalysis Letters*, Vol. 31, pp. 267-272, 1995.
14. Z. Luan, C. F. Cheng, W. Zhou and J. Klinowski, "Mesopore Molecular Sieve MCM-41 containing framework Aluminum," *in Journal of Physical Chemistry*, Vol. 99, pp. 1018-1024, 1995.
15. M. L. Occelli, S. Biz, and A. Auroux, "Mesoporous Solids," submitted.
16. M. L. Occelli and A. Schwitzer, unpublished results.
17. E. P. Barrett, L. G. Joyner, and P. P. Halenda, *J. Am. Chem. Soc.*, 73, 373 (1951).

18.   T. R. Hughes and H. M. White, *Journal of Physical Chemistry*, 71, 7 (1967).

19.   E. A. Paukshtis and E. N. Yurchenko, *Russian Chemical Review*, 52, 3 (1983).

20.   A. Samoson, E. Lippmaa, G. Engelhardt, D. Lohse, and H. G. Jerschwitz, *Chem. Phy. Let.*, vol. 134, p. 589.

21.   J. P. Gilson, G. C. Edwards, A. K. Peters, K. Rajagopalan, R. F. Worrnsbecher, T. G. Roberie, and M. P. Shatlock, J. Chem. Soc. Chem. Comm., p. 91 (1987).

NMR STUDIES OF MOLECULAR DIFFUSION WITHIN ZEOLITES OF THE TYPE USED IN FCC PREPARATION

M.-A. Springuel-Huet[1], A. Nosov[2], P. Ngokoli-Kekele[1], J. Kärger[3], J.M. Dereppe[4], J. Fraissard[1]

1. Laboratoire de Chimie des Surfaces, Université P. et M. Curie, 4 place Jussieu, 75252 Paris Cedex 05 FRANCE
2. Boreskov Institute, Lavrentieva 5, 630090 Novosibirsk, RUSSIA
3. Fakultät für Physik und Geowissenschaften, Universität Leipzig, Linnéstr. 5, D-04103 Leipzig, GERMANY
4. Laboratoire de Chimie Physique et de Cristallographie, Université Catholique de Louvain, B-1348 Louvain-La-Neuve, BELGIUM

ABSTRACT

The ^{129}Xe NMR spectroscopy of xenon adsorbed in H-ZSM-5 zeolite is used to study the rate of adsorption in of benzene and n-hexane from the gas phase. In case of benzene, the rate of the combined process of adsorption and desorption in a bed of H-ZSM-5 zeolite, consisting of crystallites whose one part have been initially benzene-loaded and the other part is free of benzene, has been studied. The exchange rate is shown to depend drastically on the crystallite arrangement. The magnitudes of the observed rate constants and line shifts are confirmed by calculations based on simple models. ^{1}H imaging NMR is used to study the distribution of benzene molecules along the sample during adsorption from the gas phase. The diffusion constants determined with the two techniques are in good agreement.

Thus, in octane FCC containing ZSM-5 crystals dispersed in an amorphous matrix, the distribution and the location of the H-ZMS-5 crystallites will be important to the transport properties of the solid and may control the performance of octane-enhancing additives containing zeolites.

INTRODUCTION

The use of ZSM-5-containing additives in FCC preparation has now become a favorite approach that refiners use to improve gasoline octane values (1). This zeolite selectively cracks low molecular weight olefins generated in a FCCU during gas oil cracking. Thus, by removing low octane components, gasoline octane values increase at the expense of gasoline yields (1).

The efficiency of the technical application of zeolites as catalysts (2-4) and molecular sieves (3) may depend critically on their transport properties.^{129}Xe NMR has proved to be a valuable tool for monitoring molecular distribution during the process of adsorption or desorption in assemblages of zeolite crystallites (5). Since the chemical shift is *inter alia* a monotonic function of the concentration of the co-adsorbed species (6), the ^{129}Xe NMR spectrum reflects the distribution of concentrations of the coadsorbed species within the sample.

In previous studies the rate of molecular propagation both through beds of zeolite crystallites (7-9) and in the intracrystalline (10) space has been investigated by ^{129}Xe NMR.

^{1}H NMR imaging allows direct observation of the distribution functions of molecules containing hydrogen atoms in the direction along which a field gradient is applied (11). It is then possible to study the kinetics of physical adsorption (12)

In the present paper the intracristalline diffusion and the influence of crystallite mixing under non-equilibrium conditions are studied and compared to ^{1}H imaging NMR experiments.

EXPERIMENTAL

Sample Preparation and Design of Sorption Experiments.

^{129}Xe NMR experiments

Zeolite crystallites of the H$^+$-ZSM-5, with a Si/Al ratio of 40, were used. Microscope observation shows that roughly a third of the crystallites were about 30 x 30 x 100 μm^3 in size; the rest being 30 μm diameter spheres (Fig. 1). Benzene (from Prolabo) and n-hexane (from Merck) were 99.7% and 99.0% pure, respectively. All experiments were carried out under natural xenon, i.e. containing 26.44% of the ^{129}Xe isotope (from Air Liquide), at 31.6 kPa (240 Torr) and 300 K. Under these conditions, the sorbate concentration was about 8 Xe atoms per ZSM-5 unit cell. Three different types of adsorption/desorption experiments were considered.

(i) *Adsorption from the gas phase*. The actived ZSM-5 zeolite placed in a glass tube in contact with xenon was separated from the gas under study (benzene or n-hexane) by a valve at a height of about 30 cm from the sample. The total amount in the gas phase corresponded to a sorbate concentration of 6 molecules per ZSM-5 unit cell at adsorption equilibrium. The process of adsorption was initiated by opening the valve.

(ii) *Sorbate equilibrium between benzene-loaded and unloaded beds of zeolite crystallites*. A fraction of benzene-loaded ZSM-5 zeolite [obtained after procedure (i)] was carefully poured over a layer of crystallites of the same zeolite at the same xenon pressure, but without the adsorbate. Benzene desorption from the upper layer and its adsorption by the lower layer was monitored by ^{129}Xe NMR.

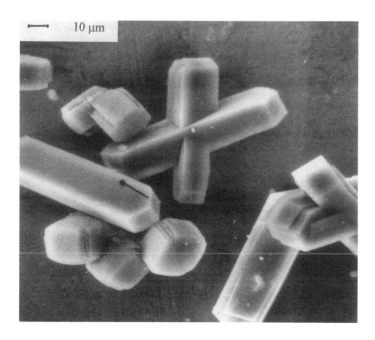

Figure 1: Electron micrograph of the H-ZSM-5 sample used for ^{129}Xe NMR experiment, magnification 600 times.

(iii) *Sorbate equilibrium between well-mixed benzene-loaded and unloaded crystallites.* As in procedure (ii), identical fractions of loaded and unloaded crystallites were brought together and thoroughly mixed, all operations being conducted under xenon. The tube containing the mixture was then introduced in the NMR probe.

<u>^{1}H imaging NMR experiments</u>

Zeolite crystallites of the H^+-ZSM-5 from Süd Chemie (Germany) with a Si/Al ratio of 130 were used. The crystallites were almost spherical with a diameter of about 0.1 μm and formed aggregates of about 20 to 50 μm. The adsorption of benzene was studied at 300 K in the absence of xenon gas and according to the general procedure (i) (see above), except that benzene vapour was used in equilibrium with liquid benzene. Therefore it could be assumed that the adsorption of benzene was conducted at a constant pressure of ca. 92 Torr (saturation vapour pressure).

NMR Spectroscopy.

The extrapolated shift of bulk xenon at zero density is used as the reference for these measurements. All resonances downfield of this reference are considered to be positive. ^{129}Xe NMR measurements were performed on a 400 MHz MSL Bruker spectrometer at 110.64 MHz. A pulse duration of 2 microsecondes and a repetition time of 0.1 s were chosen to ensure a maximum signal-to-noise ratio during time intervals which are small compared to the exchange times.

^{1}H imaging experiments were performed on a Bruker spectrometer at 300 MHz. The field gradient applied in the direction of the axis of the tube containing the sample (which was also the direction of the external field) was 47 μtesla/mm. The signal is the Fourier transform of the spin echo obtained after a conventional $\pi/2\text{-}\tau\text{-}\pi\text{-}\tau$ pulse sequence. The field gradient is applied twice, before the π refocussing pulse and during the acquisition of the signal. The $\pi/2$ pulse and the gradient durations were 12.8 μs and 10 μs, respectively and the repetition time 1 s. The time interval between the two gradient pulses was 2.5 ms. 1000 scans were run in order to obtain a satisfactory signal-to-noise ratio.

RESULTS

Figure 2 shows the ^{129}Xe NMR spectra during the adsorption of n-hexane (Fig. 2a) and benzene (Fig. 2b) by a bed of ZSM-5 zeolite [procedure (i)]. The spectra consist of two lines. The line at lower chemical shift, line a, may be attributed to xenon atoms in adsorbate-free zeolite crystallites. The second line, b, much broader for benzene than for n-hexane, corresponds to Xe atoms in loaded crystallites. The onset of this second line is observed about 1 hour (a little less for n-hexane) after valve opening. It is accompanied by a small displacement of line a to higher chemical shifts.

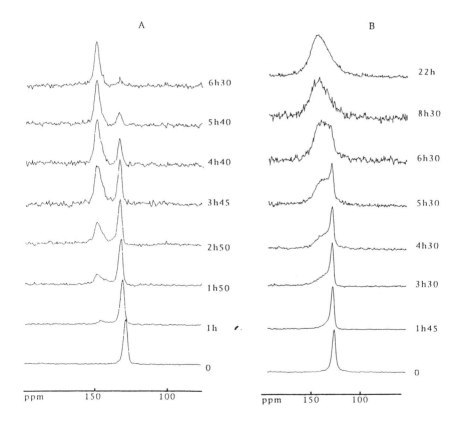

Figure 2: Variation of the ^{129}Xe NMR spectrum of a bed of ZSM-5 zeolite crystallites during the uptake of n-hexane (A) and benzene (B) from the gas phase [procedure (i)]. Time is expressed in hour and minutes.

Figure 3 shows the spectra recorded during transport of benzene from the loaded layer to an initially unloaded layer of ZSM-5 according to procedure (ii). This process is found to be slightly faster than the uptake experiment (Figure 2b). Equilibrium is achieved in about 7.5 hours. Moreover, the position of the narrow line appears to be constant throughout the experiment.

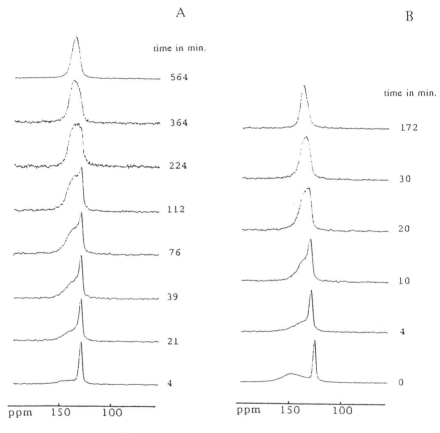

Figure 3. Variation of the [129]Xe NMR spectrum during adsorbate equilibration between an unloaded and a benzene-loaded layer of ZSM-5 [procedure (ii)] (A) and when the initially unloaded and loaded crystallites are intimately mixed [procedure (iii)] (B) (from Ref. 14, with permission).

When desorption from the loaded and adsorption on the unloaded zeolite crystallites is preceded by vigorous shaking [procedure (iii)], the adsorption equilibrium is reached after a much shorter time (Fig. 3b). One hour after the onset of the experiment, the [129]Xe NMR spectrum has already attained its final shape. The spectrum at time zero is the superposition of the spectra of the samples with and without benzene, which were measured separately before the loaded and unloaded crystallites were mixed.

Figure 4 shows the time dependence of the echo profiles. The variation of the intensity (relative to the maximum intensity) of these profiles, expressed in terms of the square root of time, is given in Figure 5.

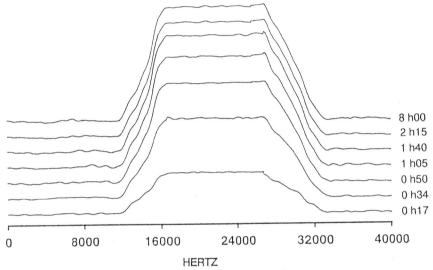

Figure 4: Time-dependence of the echo profile (^{1}H NMR imaging).

DISCUSSION

^{129}Xe NMR experiments

It has been shown that during the adsorption/desorption process of the thoroughly mixed zeolite crystallites the ^{129}Xe NMR spectra indicate that the adsorption/desorption process is controlled by intracrystalline diffusion (10). In each crystallite the concentration gradient of benzene can be expressed by (13):

$$\frac{c(r,t)}{c_\infty} = 1 \pm \frac{2R}{\pi r} \sum_{n=1}^{\infty} \frac{(-1)^n}{n} \sin\frac{n\pi r}{R} e^{\frac{-Dn^2\pi^2 t}{R^2}} \qquad [1]$$

where the plus and minus signs refer to the case of adsorption and desorption, respectively, and where the crystallites are assumed to be spheres of radius R.

The simulation of the spectra using these concentration in function of time allowed to determine a diffusion coefficient $D = D_{intra} = 1.3 \ 10^{-14}$ m^2s^{-1} (10).

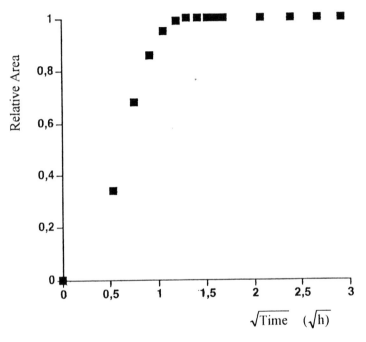

Figure 5: Variation of the relative intensity of the profiles with the square root of time.

When the loaded and unloaded crystallites are contained in two separate layers [procedure (ii)], the rate of attaining macroscopic equilibrium is controlled by the rate of molecular propagation through the bed. As a rough estimate of the time constant for attaining equilibrium we apply the relationship of the first moment t_{bed}, of the adsorption/desorption curve on a parallel-sided slab of thickness, ℓ, with the lower face isolated and with the upper face kept at a roughly constant concentration (14):

$$t_{bed} = \frac{\ell^2}{3D}$$ [2]

where D stands for the diffusion coefficient within the slab. In the present context, D is determined by the rate of molecular propagation through the bed and is given by the relationship (3):

$$D = \frac{\partial \ln P(c)}{\partial \ln c} p_{inter} D_{inter} \qquad [3]$$

where p_{inter} and D_{inter} denote the relative number of benzene (or n-hexane) molecules within the intercrystalline space and their diffusivity, respectively. The "thermodynamic factor", $\frac{\partial \ln P(c)}{\partial \ln c}$, with P(c) denoting the adsorbate pressure in equilibrium with the adsorbate concentration, c, takes into consideration the fact that the observed molecular transport occurs under non-equilibrium conditions (2).

The coefficient of intercrystalline diffusion, D_{inter}, has been estimated to 2.10^{-5} m^2s^{-1} for benzene, using the gas kinetic approach and taking into account the presence of the crystallites; these enhance the diffusion paths as compared to the gas phase (14). For benzene the quantities p_{inter} and $\frac{\partial \ln P(c)}{\partial \ln c}$ have been estimated of about 4.10^{-5} and 1.6, respectively (14).

With this estimate, the diffusion coefficient D through the bed is found to be $1.3.10^{-9}$ m^2s^{-1}, and from Equation 2 the time constant, t_{bed}, of the desorption/adsorption process is therefore 7 h (for $\ell = 10$ mm); this is in fact the order of magnitude observed in the experiments.

The retardation of uptake from the gas phase [procedure (i)] in comparison with the adsorption/desorption exchange between the two layers [procedure (ii)] is a consequence of the spatial extension of the gas volume over the bed of zeolite crystallites. Since the gas volume containing the adsorbate is separated from the upper face of the layer by a distance of 30 cm, there is clearly some delay between the instant when the valve is opened and the onset of adsorption in the layer. Equation 2 may be used to estimate this delay. Using the values obtained by the kinetic approach, D_{gas} (benzene) = $1.6.10^{-5}$ m^2s^{-1} and D_{gas} (hexane) = 3.10^{-5} m^2s^{-1}, one obtains a time lag of about 30 and 15 mn respectively, which are of the same order of magnitude as the experimentally observed delays.

In contrast to the situation in procedure (ii), in procedure (i) the concentration at the upper face of the layer is not constant but is obviously determined by the competing processes of sorbate propagation to lower layers and benzene (or n-hexane) supply from the gas phase. In this case the equation 2 should contain an additional term which has been shown to be of the order of magnitude of the time lag in the uptake experiment (14).

In procedure (i), especially in the case of n-hexane for which the lines are narrow, it is clearly seen that the intensity of line a decreases as that of line b increases. This is consistent with a propagating adsorption front where the adsorbate concentration is either zero (ahead of the front) or equal to the saturation concentration (behind the front).

The appearance of the second line in the ^{129}Xe NMR spectrum is accompanied by a shift of the line a to higher chemical shifts (Fig. 2), while in procedure (ii) the chemical shift of this initial line remains constant (Fig. 3a). In procedure(ii) the adsorption of benzene by one fraction of the crystallites is accompanied by a desorption of benzene from the other fraction. Hence, the overall xenon pressure in the sample remains roughly constant. In the case of procedure (i) benzene or n-hexane adsorption from the gas phase initiates desorption of the xenon atoms, causing the xenon pressure in the gas phase to increase and, therefore, an instantaneous increase in the xenon concentration in the crystallites without adsorbate and a corresponding downfield displacement of the line a. This process induces a small shift of the two lines as adsorption progresses.

^{1}H NMR imaging

The echo profiles reflect the time-dependence of the distribution of the benzene molecules along the sample at least 14 mm after the valve has been opened. Without xenon in the gas phase, and at rather low pressure, benzene molecules diffuse very rapidly in the entire bed. The increase in intensity of the profile corresponds to the increase in benzene concentration inside the crystallites. Assuming that diffusion in macropores and benzene vaporization are much faster than intracrystallite diffusion, we can consider that the benzene pressure is constant in the sample. Therefore the ratio of the quantity adsorbed at time t to the quantity adsorbed at equilibrium should depend on the square root of t for small values of t (15):

$$\frac{A_t}{A_\infty} = \frac{6}{R} \sqrt{\frac{Dt}{\pi}}$$ [4]

where A_t and A_∞ are the areas of the line at time t and at equilibrium, D the diffusion constant and R the radius of the spherical crystallites. It can be seen that the areas are proportional not only to the quantity of benzene inside the crystallites but also to that in the macropores. However, a rapid calculation shows that the amount of benzene in the intercrystallite space is quite negligible even for the first spectrum recorded at t = 17 mn.

Taking for R the value of the average radius of the agggregates (ca. 18 μm), the slope of the linear part of the plot A_t / A_∞ = f($\sqrt{t}$) (Fig. 5) leads to a diffusion coefficient of 9.10^{-15} m^2s^{-1} which is not far from the value determined by ^{129}Xe NMR.

We can see that the beginning of the curve A_t / A_∞ = f($\sqrt{t}$) is not linear, the curve being somewhat sigmoid. This deviation has often been observed and is attributed to a variety of causes, such as gas-phase film mass transfer or surface barriers (structural barrier, evaporation barrier, thermal barrier, etc.)

CONCLUSION

^{129}Xe NMR demonstrates the dramatic effect of the arrangement of crystallites on the rate of intercrystalline exchange of guest molecules in zeolite beds when the adsorbate pressure is small. The exchange rate of benzene molecules in a bed of well mixed crystallites of ZSM-5 zeolite is at least one order of magnitude larger than the corresponding molecular exchange between two separate layers of crystallites, even for small bed heights. It is controlled in this case by intracrystalline diffusion. Thus HZSM-5 crystallites location and distribution will be critical to the performance of octane enhancing additives containing this type of zeolite. The observed rate constants for the uptake from the gas phase [procedure (i)] and for molecular exchange between two separate layers [procedure (ii)] are consistent with simple models. The displacement of the ^{129}Xe NMR line corresponding to adsorbate-free crystallites during adsorbate uptake [procedure (i)] could be attributed to the enhancement of the xenon pressure in the gas phase.

^{1}H NMR imaging is a particularly interesting technique for visualizing the distribution of the adsorbate in a sample. From the variation of the profiles with time, we have been able to determine a diffusion coefficient in good agreement with that obtained by ^{129}Xe NMR.

References

1. Schipper, P. H., Dwyer, F. G., Sparrel, P. T., Mizrahi, S. and Herbst, A. in " *Fluid Catalytic Cracking: Rôle in Modern Refining* " M. L. Occelli (ed.), ACS Symposium Series n° 375, p. 64 (1988)
2. Satterfield, C. N., *Mass Transfer in Heterogeneous Catalysis*, M.I.T. Press: Cambridge, MA. 1970.
3. Kärger, J. and Ruthven, D.M., "*Diffusion in Zeolites and Other Microporous Solids*", Wiley: New York, 1992.
4. Chen N. Y., Degnan, T. F and Smith, C. M., "*Molecular Transport and Reaction in Zeolites*", VCH: New York, 1994.
5. Fraissard, J. and Ito, T., *Zeolites*, , 8, 350 (1988)
6. Gedeon, A., Ito, T. and Fraissard, J., *Zeolites*, 8, 376 (1988)
7. Chmelka, B. F., Gillis, J. V., Petersen, E. E. and Radke, C. J., *AIChE-J.*, 36, 1562 (1990)
8. Bansal, N. and Dybowski, C. J., *J. Magn. Reson.*, 89, 21 (1990)
9. Ryoo, R., Kwak, J. H. and de Menorval, L. C., *J. Phys. Chem.*, 98, 7101 (1994)
10. Kärger, J., Pfeifer, H., Wutscherk, T., Ernst, S., Weitkamp, J. and Fraissard, J., *J. Phys. Chem.*, 96, 5059 (1992)
11. Callagan, P. T., "*Principles of Nuclear Magnetic Resonance Microscopy*, Oxford University Press: London 1991
12. Heink, W., Kärger, J., Pfeifer, H., *Chem. Eng. Sci.*, 33, 1019 (1978)
13. Crank, J., "*The Mathematics of Diffusion*" Clarendon Press: Oxford, 1956
14. Springuel- Huet, M.-A., Nosov, A., Kärger, J. and Fraissard, J., *J. Phys. Chem.*, 100, 7200 (1996)
15. Barrer, R. M. and Clarke, D. J., *J. Chem. Soc., Faraday Trans. I*, 70, 535 (1974)

THE USE OF MICROCALORIMETRY AND POROSIMETRY TO INVESTIGATE THE EFFECTS OF AGING ON THE ACIDITY OF FLUID CRACKING CATALYSTS (FCC)

M. L. Occelli(1), A. Auroux(2), F. Baldiraghi(3), and S. Leoncini(3)

1. GTRI, Georgia Institute of Technology, Atlanta, Georgia 30332 USA
2. CNRS, Villeubanne, FRANCE
3. EURON, San Donato Milanese, Milan, ITALY

Keywords: microcalorimetry, probe molecules, steam aging, and metal
             contaminants

I. ABSTRACT

Microcalorimetry experiments, with ammonia and pyridine as probe molecules, have been used to investigate the effects of thermal and hydrothermal treatments on the acidity of a commercially available fluid cracking catalyst (FCC). Results have been compared with those generated from an equilibrium sample of the same catalyst obtained from a European refinery. Variations in the FCC internal structure was instead studied using pore size distribution data from isotherms calculated by density functional theory (DFT) method and by using the Barrett, Joyner, and Halenda (BJH) equation.

Pore size distribution results indicate that aging decrease the FCC microporosity and change the relative distribution of pore sizes in the meso-macro porosity range. As a result, the sorption capacity for the probe molecules used, as well as the FCC acid site density, are drastically reduced. The acid site density and strength of the equilibrium FCC does not deviate significantly from that of the steam-aged catalyst indicating that the steaming procedure used reduced the acidic properties of the fresh catalyst to those of the corresponding equilibrium sample.

The difference in cracking activity between two aged FCC having similar acidic properties illustrate the importance that the catalyst internal and surface architecture, metal impurities levels and cracking sites availability have on FCC performance.

II. INTRODUCTION

Theoretical efforts are becoming increasingly important in catalysis. Theory provides a framework for understanding the relationships among catalyst composition, structure and performance, and when judiciously applied, can greatly reduce experimentation.   However, the industrial development of catalysts is and will remain dominated largely by experimental studies because catalysts cannot yet be designated from first principles.   Formulation, characterization, microactivity testing, pilot plant testing and product analysis cannot be avoided. The use of MAT to evaluate FCC is probably the best and most easily recognized example of the usefulness of microreactors for the bench-scale evaluation of catalysts.

Catalyst study and development should conclude with pilot plant testing of candidates identified by MAT evaluation.   The development of a correlation between theoretical calculations, physicochemical properties, MAT results, pilot plant data and data from a Fluid Cracking Catalyst Unit (FCCU) in a refinery is what justifies, in part, the high cost of research in catalysis.

MAT results and their utility, greatly depend on the steam aging procedure used to reduce the physiocochemical properties of newly synthesized FCC to those of the corresponding equilibrium material.   Steam-aging (at 815°C/5 h with 100% steam, 1 atm) is a procedure thought capable of reducing the physical properties (surface area, pore volume and crystallinity) of certain fresh FCC to those measured in equilibrium samples.   These results explain only in part the similarities and differences in MAT results (during gas oil cracking) obtained with steamed and equilibrium FCC.

It is the purpose of this paper to report the use of microcalorimetry and porosimetry experiments to investigate the effects of aging (at the aforementioned MAT conditions) on FCC acidic and structural properties (1).

Although adsorption calorimeters are generally used at temperatures lower than those used in most hydrocarbon conversion reactions, heat flow calorimetry appears to be one of the most suitable and accurate techniques to study the number and strength of acid sites in FCC by adsorption of basic probe molecules. Calorimetry allows not only a study of acid sites distribution but also, at least in favorable cases, a study of reaction mechanisms and kinetics. The technique and the determination of acid-base properties of molecular sieves by adsorption calorimetry have been reviewed in comprehensive articles by Cardona-Martinez and J. Dumesic (2), P. Anderson and H. Kung (3), W. Farneth and R. Gorte (4), and Auroux (5,6).

III. EXPERIMENTAL

Surface area and pore volume of the fresh, steam-aged and equilibrium FCC were measured with an ASAP-2000 porosimeter from Micromeritics, see Table 1. Prior to nitrogen adsorption, the samples were degassed in vacuum at 400°C/4h. Pore size distribution in the mesoporous region was obtained with the BJH equation (7) and using nitrogen sorption isotherms obtained from Pi/P values > 0.01. The microporous region was instead investigated using DFT methods (8).

Heat of adsorption of NH_3 and pyridine was measured using a heat-flow microcalorimeter of the Tian-Calvet type linked to a glass volumetric line. Successive doses of gas were sent onto the sample until a final equilibrium pressure of 133 Pa was obtained. The equilibrium pressure relative to each adsorbed amount, was measured by means of a differential pressure gauge from

Table 1. Physical Properties and Composition Data of the Commercial FCC
Samples Under Study.

	Fresh	Steamed	Equilibrium
Total Surface Area (m^2/g)	334	175	120
Micropore Surface Area (m^2/g)	236	122	88
Pore Volume (ccH$_2$O/g)	0.48	0.42	0.38
PV (ccN$_2$/g)	0.20	0.17	0.13
APD (4V/A, BET) nm	2.3	4.0	—
APD (BJH) nm	5.4	11.2	14.9
A. B. Density (g/cc)	0.72	0.76	0.82
Al$_2$O$_3$ (wt%)	48.0	48.2	41.0
Re$_2$O$_3$ (wt%)	1.78	1.82	1.79
Na$_2$O (wt%)	0.23	0.23	—
Fe (wt%)	—	—	0.53
V (wt%)	—	—	0.46
Ni (wt%)	—	—	0.15

Datametrics. The adsorption temperature was maintained at 100°C. Primary
and secondary isotherms were collected at these temperatures. All samples were
dried overnight under vacuum at 400°C before calorimetric measurements were
undertaken.

VGO was used to perform MAT evaluation of the various FCC. A
cat/oil ratio of 4, a 30 sec injection time and a reactor temperature of 527°C
were used. Prior to MAT analysis, all FCC were steam aged at 815°C/5h with
100% steam at 1 atm. All powder diffraction measurements were obtained with
a Rigaku diffractometer at a scan rate of 1°/min using monochromatic Cu-kα
radiation.

IV. RESULTS AND DISCUSSION

Aging of fresh FCC results in crystallinity lossess of the type shown
in Fig. 1. These diffractograms contains variation in peak positions, peak

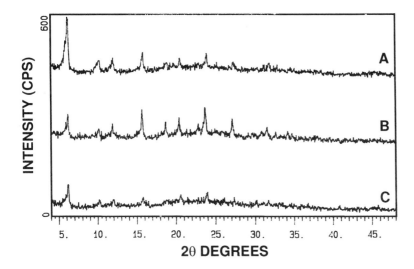

Figure 1. XRD diffractogram of: A) Fresh FCC, B) Steam aged FCC, and C)
Equilibrium FCC.

broadening and reductions in peak intensity consistent with a decrease in crystallinity of the HY type zeolite present in these catalysts.   The enhanced decrease of crystallinity in the equilibrium FCC shown in Fig. 1C, is attributed to the presence of 0.46% V on this sample, see Table 1.

The determination of pore sizes distribution in the micropore (i.e., pore diameter < 2 nm) region is still a subject of controversy (7-15). Of particular interest to FCC studies are recent methods by which model isotherms, calculated from DFT, are used to determine pore size distribution from experimental isotherms (8).   The method assumes a slit-type geometry and is applicable over the entire range of pore sizes available to the probe molecule (8). AFM results have indicated that slit-type geometry is fairly common on the FCC surface (16).

The effects of the hydrothermal procedure used on the fresh FCC porous structure, are shown in Table 1 and Figs. 2-3. Steaming decreases the fresh FCC microporosity (Table 1). These changes are even more severe in the equilibrium sample, Fig. 2. The decrease in microporosity in Fig. 2 is attributed to crystallinity losses of the FCC cracking component (a HY-type zeolite). The large reduction in microporosity observed in the equilibrium sample is the result of an enhanced decrease in crystallinity due to the presence of 0.46% V; see Table 1 and Fig. 2. Aging affect also the samples macroporosity, Fig 2. Changes in the meso and macroporous range are better observed using the BJH equation (7), Fig 3.

Steam-aging seems to preferentially collapse the FCC small mesopores thus increasing the average pore diameter of a residual structure that has suffered

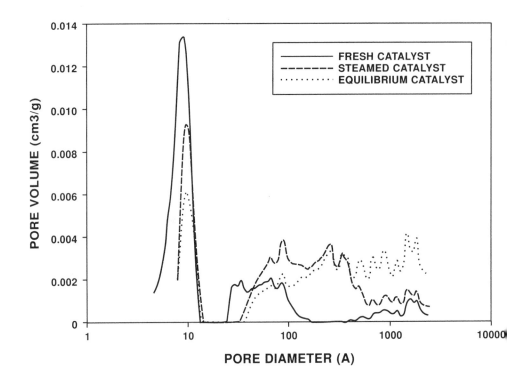

Figure 2. Pore size distribution from Density Functional Theory (DFT) methods (9).

a reduction in both surface area and total pore volume, see Table 1. Initially, the fresh FCC contains a bimodal distribution of pores with diameter in the 2.0 to 10.0 nm range, Figs. 2,3. The collapse of the mesopores with APD near 2.6 nm upon steaming, results in the formation of some macropores with APD in the 20.0 to 50.0 nm range, Fig. 3. Metals oxides (NiO and V_2O_5) could have caused the blockage of the mesopores with APD near 10.0 nm, Fig. 3. The blockage of this type of pores (surface cracks and slits) by vanadia has been inferred also by AFM images of a V-contaminated FCC (16). Metals (Ni, Fe, and V) deposition in the refinery FCCU, is believed responsible for the equilibrium FCC sample greater losses in physical properties and crystallinity shown in Table 1 and Figs. 1-3.

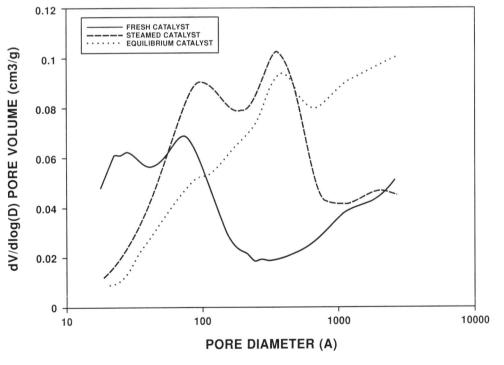

Figure 3. Pore size distribution in the meso and macroporous range from the BJH equation (8).

Heat flow microcalorimetry has become a well established technique to measure the total acidity of catalysts (17-19). By recording differential heats of sorption of probe molecules such as ammonia and pyridine, it is possible to obtain information on the solid acid site density and strength. The application of microcalorimetry to the study of catalysts has been reviewed in recent articles (17-19). Some difficulty was encountered in obtaining data at 150°C for equilibrium samples. It is believed that at 150°C, the metal impurities present on the catalyst surface (such as vanadia) catalyze the decomposition of the probe molecules. This problem was circumvented by repeating the experiments at 100°C.

The sorption isotherms in Fig. 4A indicates that steam aging drastically reduce the number of sites available to NH_3 chemisorption and that these sites are similar in number to the one observed in the equilibrium sample. These results are consistent with the decreased microporority of the catalysts after aging, Table 1. Differential heats of adsorption are given in Fig. 4B. At low (<50 μmol NH_3/g) coverage, the FCC contains sites with strength in the 170-180 kJ/mol range. As NH_3 coverage increases, the fresh FCC exhibit a fairly large population of sites with strength near 140 kJ/mol attributed to the presence of an estimated 20-30 wt% HY type zeolite. After sorbing about 200 μmol NH_3/g, heats of sorption monotonically decrease with coverage until a second population of sites with strength near 60 kJ/mol is observed. These weak sites are attributed to H-bonding of NH_3 with SiOH groups (1) and possibly, to weak Lewis acid centers associated with the FCC binder. Results in Fig. 4B show that the fresh FCC initial acid (B+L) sites strength decreases from 180 kJ/mol to 140 kJ/mol after steam aging, and to 125 kJ/mol after use in the refinery FCCU. The two aged FCC exhibit similar losses in acid site strength and density at higher NH_3 coverage; Fig. 4B. Results for pyridine chemisorption are shown in Fig. 5.

The sorption isotherms for pyridine are similar to the ones observed for NH_3, Fig. 5A. However in contrast to ammonia, pyridine because of its size, is

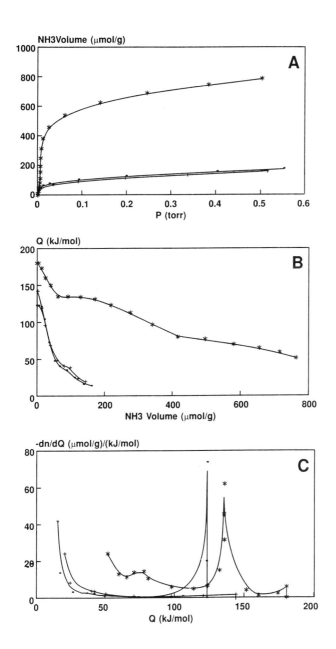

Figure 4.

Microcalorimetry results from NH3 chemisorption. A) Sorption
isotherms, B) Differential heats of adsorption, and C) Acid site
strength distribution. The symbols (-), (+) and (*) refer to:
equilibrium, steamed and fresh FCC samples.

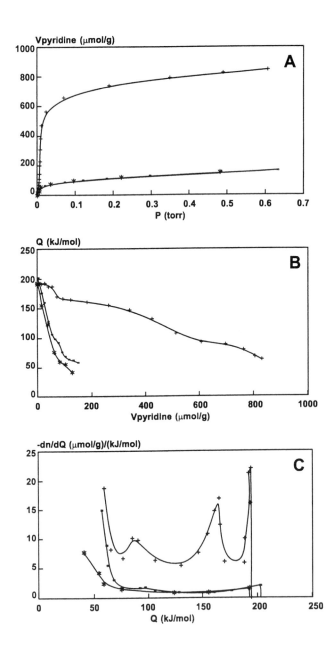

Figure 5.

Microcalorimetry results from pyridine chemisorption. A) Sorption isotherms, B) Differential heats of adsorption, and C) Acid site strength distribution. The symbols (∎), (+) and (∗) refer to: steamed, fresh and equilibrium FCC samples.

probably preferentially sorbed on acid sites in the mesoporous structure so that initially all the three FCC samples exhibit sites with strength in the 180-200 kJ/mol range, Fig. 5B. Then, as pyridine coverage increases, the fresh FCC is found to contain a population of sites with strength near 170 kJ/mol and one near 105 kJ/mol while in the two aged FCC only an heterogeneous distribution of sites is observed, Fig. 5B. The large decrease in acid sites density observed with ammonia is also seen in the differential heat of pyridine adsorption shown in Fig. 5B. In agreement with the results of Chen et al., (1) steam aging has little effect on the FCC initial acid site strength, but drastically decrease the catalysts acid site density and distribution.

An heterogeneous distribution of sites is present in steam aged and equilibrium samples, Figs. 4B, 5B. Distribution of acid site strength is better osberved by plotting the number of sites with a given strength versus the strength of the site as in Fig. 4C. The area under the curve will then give the number of sites (the population) having a given strength.

The fresh FCC exhibit a minor population of strong acid sites with strength near 175 kJ/mol and a major one with strength near 140 kJ/mol, see Fig. 4C. The equilibrium FCC contains a small population of sites near 125 kJ/mol absent in the steam aged FCC that could be attributed to NH_3 interactions with metals contaminants. With pyridine, these populations are not seen and an heterogeneous distribution of acid sites is present in the two aged samples, Fig. 5C. In contrast,when exposed to pyridine, the fresh FCC reveals the presence of three types of sites with strength near 188, 165, and 82 kJ/mol, Fig. 5C.

Results in Figs. 4, 5 indicate that the steam aging procedure used can indeed reduce the fresh FCC total acidity to that of the corresponding equilibrium sample obtained from a major European refinery.

Microactivity results from the aged FCC samples are given in Table 2. Although these FCC have similar acidity, their cracking activity is significantly different. In Table 2, differences in activity and liquid product selectivities

Table 2. Microactivity Test Results for Steam Aged and Equilibrium FCC
Sample taken from a Major European Refinery.

	(Steamed)	(Equilibrium)
Conversion (V % ff)	72.1	62.9
Gasoline (V % ff)	49.6	41.0
LCO (V % ff)	18.4	18.9
HCO (V % ff)	9.5	18.2
L.P.G. (wt % ff)	5.9	14.2
C_3 (wt % ff)	1.1	0.9
$C_3^=$ (wt % ff)	4.7	4.4
iC_4 (wt % ff)	3.1	2.6
nC_4 (wt % ff)	0.9	0.7
$iC_4^=$ (wt % ff)	1.6	1.6
$C_4^=$ (wt % ff)	4.5	4.0
Fuel Gas (wt % ff)	3.0	2.9
Coke (wt %)	3.6	4.8

between steam aged and equilibrium FCC, are attributed mainly to crystallinity losses and to variations in the equilibrium FCC micro and mesopore structure resulting from vanadia (and nickel) deposition during operation in the refinery FCCU (20-22).

The presence of Ni (and Fe) impurities on the equilibrium FCC surface promotes secondary cracking reactions of hydrocarbons in the gasoline boiling range thus contributing to the higher than expected LPG and coke yields reported in Table 2.

SUMMARY AND CONCLUSION

Although in a typical commercial fluid cracking unit steam stripping of occluded hydrocarbons from the catalyst surface is performed at temperatures in the 480-540°C range, more severe hydrothermal treatments are necessary to reduce the structural and catalytic properties of certain fresh commercial catalysts to equilibrium levels in a short (5-10 h) period of time (21,22). Since different catalysts may deactivate differently, a variety of deactivation conditions are probably needed to predict the activity of each catalyst in the field.

The steam aging procedure used in the present research (at 815°C/5 h with 100% steam at 1 atm) has been found capable of reducing the acidity properties of the fresh FCC under study to those that the same FCC exhibit after use in an FCCU of a major European refinery.

Differences in activity and liquid product selectivities between the two aged FCC in Table 2, can be attributed to the presence of metal contaminants found on the equilibrium FCC surface and to the effects that these metals have on the FCC crystallinity, micro and mesoporous structure, Figures 2,3.

The overall low acidity of the steam-aged and equilibrium samples (Figures 4B,5B) is somewhat surprising. Since these catalysts retain most of their useful cracking activity (Table 2), it is believed that only acid sites (cracking centers) near the surface effectively contribute to gas oil cracking.

Surface acidity together with surface composition and a pore architecture that is open to sorption and catalysis, is probably what controls the FCC activity and selectivity properties.

V. REFERENCES

1. Chen, D. A., S. Sharma, N. Cardona-Martinez, J. A. Dumesic, V. A. Bell, G. D. Hodge, and R. J. Madon, *J. Catalysis* 136, p. 392 (1992)
2. Cardona-Martinez, N. and Dumesic, J. A., *Adv. Catal*, 38 49 (1992)
3. Andersen, P. J. and Kung. H. H., *Catalysts*, Royal Society of Chemistry, vol. 11, p. 441 (1995)
4. Farneth, W. E. and Gorte, R. J., *Chem. Rev.*, 95, 615 (1995)
5. Auroux, A., *Catalyst Characterization: Physical Techniques for Solid Materials*, B. Imelik and J. C. Vedrine Eds., Plenum Press, New York, p. 611, (1994)
6. Auroux, A., *Topics in Catalysis*, (1996) (to be published)
7. Barrett, E. P., Joyner, L. G., and Halenda, P. H., *J. Am. Chem. Soc.* 73, 373 (1951).
8. Olivier, J. P., and Conklin, W. B., at the *International Symposium on the Effects of Surface Heterogeneity in Adsorption and Catalysis on Solids*, Kazimierz Dolny, Poland, July 1992.

9. deBoer, J. H., Linsen, B. G., and Onsinga, Th. J., *J. Catalysis* 4, 3 (1965).

10. Dubinin, M. M., and Radushkevich, L. V., Doklady Akad, Nauk, S.S.S.R., 55 327 (1947).

11. Lippens, B. C., and deBoer, J. H.., J. Catalysis 4, 319 (1965) K. S. W. Sing, *Proc. Int. Symposium on Surface Area Determination*, Everett, D. H., Ottwill, R. H., eds., Butterwork, London, 25 (1970)7

12. Brunauer, S., Mikhail, R. Sh, and Bodor, E. E., J. *Colloid Interf. Sci.* 24, 451 (1967).

13. Horvath, G., and Kawazoe, K. J. *Chem. Eng.* Japan 6, 470.

14. Saito, A., and Foley, H. C., *AIChE Journal* 37, 429 (1991)

15. Seaton, N. A. Walton, J. P. R. B., and Quirke, N., *Carbon* 27, 853 (1989)

16. Occelli, M. L., and S. A. C. Gould, *CHEMTECH* 24, p. 24 (1994).

17. Auroux, A., and Ben Taarit, Y., *Therm. Acta*, 1, 63 (1987)

18. Cardona-Martinez, N. and J. Dumesic *Adv. Catal.* 38, 9 (1992)

19. Auroux, A., *Catalyst Characterization: Physical Technique for Solid Materials*, B. Imelik and J. C. Vedrine, eds.; Plenum Press, Chapter 22, p. 611 (1994)

20. Occelli, M. L., S. A. A. C. Gould, F. Baldiraghi, and S. Leoncini, this "Preprints", Volume (1996)

21. Magee, J. S.; Blazek, J. J., *ACS Monogr*, 171, 615 (1976)

22. Marshall, S. *Pet. Refiner*, 31(9), 263, (1952)

METHODS FOR THE CHARACTERIZATION OF
ACID SITES IN FCC CATALYSTS

R.J. Gorte
Department of Chemical Engineering
University of Pennsylvania
Philadelphia, PA 19104

and

A.I. Biaglow
Department of Chemistry
United States Military Academy
West Point, NY 10996

I. ABSTRACT

Questions concerning how to characterize and describe FCC catalysts will be discussed. For determining Brønsted acid site densities, it will be demonstrated that Temperature Programmed Desorption (TPD) of simple amines is very useful for the solid acids typically used in FCC catalysis. The ammonium ions formed by the interaction of amines with the acidic protons decompose to olefins and ammonia in a well-defined temperature range which allows the sites to be easily counted. Some other methods of measuring site densities are not reliable. For example, the site density is not equal to the framework Al content in faujasites and only a small fraction of the Al in amorphous silica-aluminas result in Brønsted acidity. However, characterization of acid strength is much more difficult. Insights into this question can be obtained from measurements of olefin oligimerization and n-hexane cracking, when the rates are normalized to the Brønsted site density.

II. INTRODUCTION

Despite the importance of fluid-catalytic-cracking (FCC) catalysis, there is still a great deal of confusion over even the most fundamental aspects of these reactions and catalysts. While there now appears to be a general consensus that Brønsted-acid sites are involved,[1] there is no general agreement on how one should characterize and describe the strength of the sites, whether or not there is a distribution of acid strengths in these materials, whether all protonic sites are involved in reaction, and what role Lewis sites play in the reaction.

There are a number of reasons for the confusion concerning FCC catalysis in the literature. First, the catalysts are very complex. Usually, the catalysts are a mixture of faujasite and amorphous silica-alumina materials, with additional additives such as H-ZSM-5 included for fine tuning the desired properties. In the case of faujasite materials, there may be rare-earth cations or intentional addition of alkalies. Even for "simple" H-Y, there is always a mixture of framework and nonframework Al, both of which affect reactions. Compared to amorphous silica-aluminas, the faujasite phase is easy to describe. Relatively little is known about the amorphous component, except that there are again Brønsted and Lewis sites based on pyridine adsorption measurements.

Second, it is easy to demonstrate that some of the most commonly used methods for characterization of acidity in solid acids do not work. For example, temperature-programmed desorption (TPD) of ammonia or pyridine is one of the most widely used methods for determining both the acid-site concentration and strength, yet recent work has shown that neither property can be easily determined from TPD. Juskelis and coworkers recently showed that ammonia desorbs in a similar temperature range from CaO and US-Y, which raises important questions about what sites are counted in TPD.[2] Peak desorption temperatures, a common measure of site strength, are even more misleading.[3] In a recent demonstration on H-ZSM-5, it was shown that the peak temperature for ammonia desorption changed by more than 150K depending on how the measurement was made.[4]

Third, there is no adequate scale which can be used to describe acidity in solid acids, comparable to the pKa scale for aqueous acids.[5] That solution phase scales, such as Hammett acidities, are inappropriate for solid acids is easy to demonstrate from the simple fact that pyridine, a weak base in aqueous solutions, binds much more strongly to sites in zeolites than does ammonia, which is a strong base in aqueous solutions.[6] Certainly, one cannot use the Hammett acidity to predict how a given material will interact with reactant molecules. Furthermore, it has recently been shown that two materials with very different catalytic activities for hydrocarbon reactions, H-[Al]-ZSM-5 and H-[Fe]-ZSM-5, both having only Brønsted sites, exhibited identical differential heats of adsorption for both ammonia and pyridine.[7] The lack of an adequate acidity scale and the lack of a correlation between catalytic activity and heats of adsorption for ammonia and pyridine raise serious questions about how one should define and measure "acidity" in solid acids, and about the meaning of the results from the many studies which have been carried out to determine the acid strengths and acid site distributions in solid acids.

Fourth, hydrocarbon cracking reactions are very complex. Under most conditions, the cracking of alkanes involves a chain reaction in which the saturated molecule becomes a carbenium ion through removal of a hydride ion, the carbenium ion undergoes cracking to olefin products and a smaller carbenium ion, and finally the chain is transferred to another alkane through hydride transfer to the small carbenium ion. Even if the Brønsted sites are primarily responsible for forming and maintaining the carbenium ions, it should be expected that presence of Lewis sites and the geometry of the site itself would influence the overall rates and product distribution by modifying other aspects of the chain reaction.

In this paper, we will discuss results from our work which we believe helps address some of the questions which have been raised above. Obviously, given the complexity of the issues and the large number of papers which have already been written on the subject, this paper will likely leave more questions than it answers. Nevertheless, we hope that it provides a format for looking at the issues in a new way.

III. SITE DENSITIES

Before one can properly compare different solid acids, one must be able to determine the concentration of acid sites. Site concentrations for Brønsted acid sites are especially important because there is a clear correlation between these sites and alkane cracking. As we have already discussed, ammonia desorption is unreliable as a means for determining acid site density because it cannot distinguish between noncatalytic and catalytic materials.[2] Simple alkylamines, with the exception of methylamine, are particularly useful for probing Brønsted sites. Protonation of the amines results in the formation of an ammonium ion, which in turn decomposes to ammonia and an olefin at elevated temperatures, in a reaction similar to the Hofmann elimination reaction of quaternary ammonium salts. On H-ZSM-5, the number of amine molecules which undergo reaction in TPD was found to be identical to the Al content for a whole series of amines.[8] Furthermore, no reaction of the amines was found on Lewis acids, such as alumina, showing that only Brønsted sites are probed.

The utility of using amines for the characterization of Brønsted acidity was first demonstrated on a series of FCC catalysts synthesized by the partial crystallization of calcined kaolin.[10] The starting material contained ~11% faujasite in a clay matrix, and this material was then steamed for varying lengths of time at 1060K to simulate deactivation in the regenerator. Using isopropylamine, Brønsted sites were counted by measuring the amount of propene and ammonia which formed between 575 and 650K. As shown in Figure 1a, it was found that the site density decreased linearly with the framework Al content of the faujasite phase, as determined by XRD. Based on the slope of that line, the site density appeared to be only one third of the framework Al content. Figure 1b shows that the MAT activity also decreased

a.)

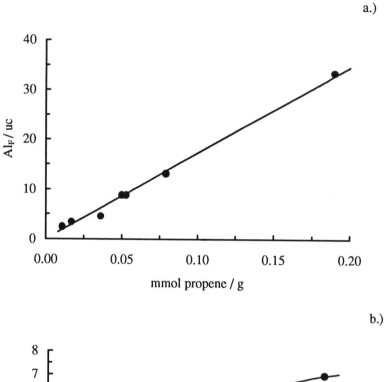

b.)

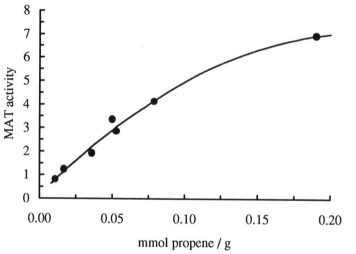

Figure 1. (a) Framework Al content as a function of Brønsted acid site
concentration, and (b) MAT activity of FCC materials.[10]

in a linear manner with the Brønsted site density, indicating that the Brønsted sites are indeed primarily responsible for the catalytic activity.

To elucidate the reason for the Brønsted site density being different from the framework Al content, we next used isopropylamine to examine a Na-Y zeolite as a function of proton exchange level.[11]   In agreement with previous IR studies, it was found that a hydroxyl peak at 3650 cm^{-1} appears first at low exchange levels, followed by a second hydroxyl feature at 3550 cm^{-1} at higher levels of exchange.   Only those isopropylamine molecules associated with the high frequency band reacted to propene and ammonia.   While evidence for interactions between isopropylamine and hydroxyls associated with the low frequency hydroxyl band was observed, both from a disappearance of the hydroxyl band in the infrared and a new, low-temperature desorption feature for isopropylamine in TPD, the molecules associated with the low-frequency band are not protonated and do not react.   Since the hydroxyls associated with the low frequency band are not easily accessible from the supercage, it appears that these hydroxyls cannot promote the reaction.

The amine desorption technique works equally well for measuring the Brønsted-acid site density in amorphous catalysts.[9]   Site densities and n-hexane cracking rates are shown in Table 1. The site density on a given

Table 1. Site concentrations and n-hexane cracking rates in silica-aluminas.[9]

Sample	Al content	[sites][a]	TOF x10^6 650K[b]
	μmol/g	μmol/g	molec./site sec.
Silica	----	5	13.4
Alumina	----	29	3.16
3.1% alumina	602	42	8.54
5.0% alumina	982	63	12.0
9.8% alumina	1910	56	11.3
13% cogel	2440	97	8.12
25% cogel	5180	108	7.38

[a]Site concentrations were determined with TPD of amines.
[b]n-Hexane cracking measurements were conducted in a differential reactor.

catalyst, measured with different amines, was found to be the same, independent of the amine chosen;[12] and, for a series of amorphous catalysts, the hexane cracking activities increased almost linearly with the site density.[9] Reaction of the amines occurs in exactly the same temperature range on the amorphous catalysts as was found in zeolites.

IV. SITE STRENGTHS

As discussed in the Introduction, this is a much more complicated concept to discuss. Solution phase concepts do not apply and there is no correlation between heats of adsorption for ammonia or pyridine and catalytic activity. In a practical sense, one would like a definition which is related to the property of interest, namely the catalytic activity for a reaction such as hexane cracking. Due to the fact that alkane cracking is a complex reaction, it is important that comparisons of rates on different materials be made for conditions in which the same mechanism applies in all cases. In alkane cracking, at least two mechanisms are distinguishable, as shown in Scheme 1 which is a unimolecular reaction, and Scheme 2, a bimolecular, chain reaction involving hydride transfer from the alkane to a carbenium ion to maintain the chain. The second mechanism, which is favored at high pressures and low

Scheme 1

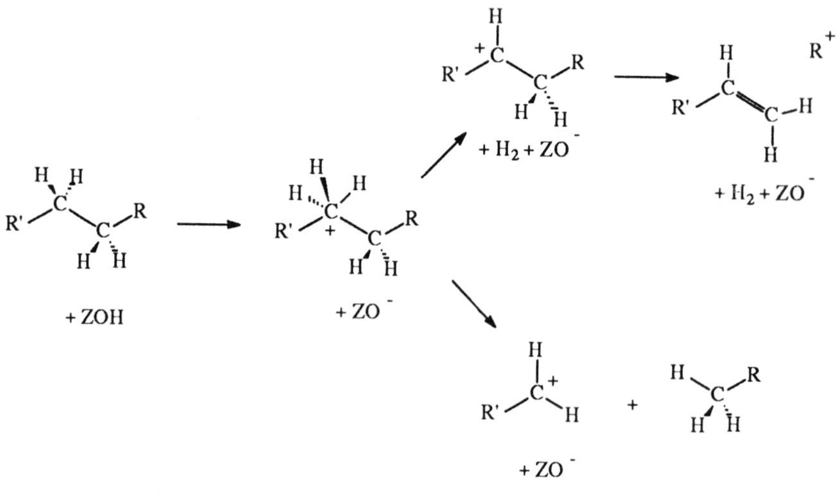

Scheme 2

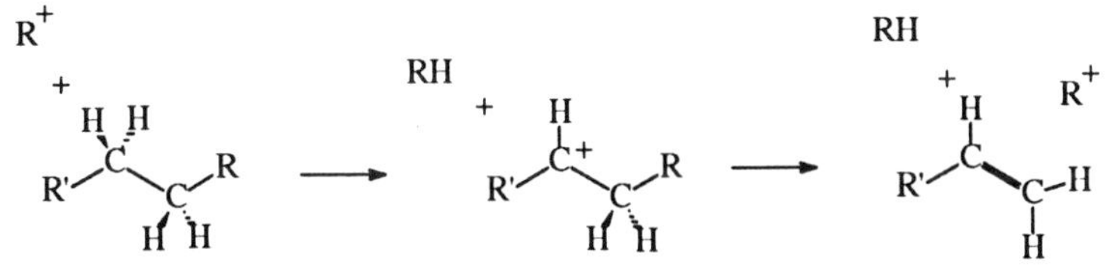

temperatures, has a distinctively lower activation energy, ~17kcal/mol, compared to almost 30 kcal/mol for the unimolecular regime.[13]

It is well known that steaming, and the resulting nonframework Al which is present following this pretreatment, can significantly increase the catalytic activity of a faujasite catalyst for alkane cracking. This is usually attributed to an increase in the "acidity" of the material. To examine this effect, we measured the cracking rates for n-hexane in a series of dealuminated faujasites in an attempt to determine how the catalytic sites are modified by steaming. The materials examined are listed in Table 2, and included the

Table 2. Site concentrations and n-hexane cracking rates in H-Y.[14]

Sample	[sites][a]	Al_f^b	Al_f^b	TOF x10^3 798K[c]	E_a
	μmol/g	atoms/u.c.	μmol/g	molec./site sec.	kcal/mol
HY(CD)	860	26.0	2260	2.5	31
HY(7)	850	25.7	2230	14	17
HY(10)	450	8.6	750	1.5	33
HY(20)	230	5.6	490	6.4	34
HY(71)	70	0.6	50	3.2	28
HY(85)	60	1.5	130	1.5	35

[a]Site concentrations determined using TPD of amines.
[b]Measured using XRD and ^{29}Si NMR.
[c]Measurements conducted in a differential flow reactor. TOF computed by normalizing reaction rates to the site concentration.

following: 1) a chemically dealuminated HY, designated as H-Y(CD), with 26 framework Al/unit cell; 2) a US-Y containing 26 framework Al/unit per; and 3) a series of H-Y catalysts, prepared by steaming and subsequent acid leaching, ranging from 9 framework Al/unit cell to <1 Al/unit cell. All samples were characterized by isopropylamine TPD, ^{29}Si NMR, and XRD. The Brønsted-site density, as determined from amine TPD, was found to depend only on the framework Al content, so that the site density for the HY(CD) sample was nearly identical to that of the US-Y, ~850 μmol/g. In order to make a fair comparison, all rates were normalized to the Brønsted site density.[14]

First, in agreement with previous studies, the rates for n-hexane cracking at 798K and 47 Torr, also shown in Table 2, were significantly higher, by a factor of ~6, on the US-Y compared to the H-Y(CD). However, this was not true for steamed samples with a lower Al content. All of these materials exhibited turnover frequencies which were practically identical to the H-Y(CD). Furthermore, the US-Y was unique in that the activation energy for cracking was 17 kcal/mol, rather than the 30 kcal/mol observed on each of the other materials. The activation energies would suggest that the bimolecular mechanism, Scheme 2, was dominant in the US-Y sample, while the unimolecular mechanism, Scheme 1, was responsible for the reaction on the other samples. This might suggest that the presence of nonframework Al, at high site densities, enhances hydride transfer rather than increasing acidity.

It is also interesting to compare results for amorphous silica-alumina catalysts to that found for the faujasites. It was reported that the average turnover frequency for a series of amorphous catalysts at 650K and 41 Torr of n-hexane was 9.3×10^{-6} sec^{-1}.[9] Using an activation energy of 30 kcal/mol and a first order dependence in pressure to extrapolate the rates to the conditions used for the faujasites, we calculate that the turnover frequency for amorphous catalysts would be ~0.7×10^{-3} sec^{-1}, compared to a rate of 2.5×10^{-3} for the H-Y(CD). This is a relatively small difference which could easily be due to sorption properties,[15] so that there is no reason to suggest that the catalytic activity of amorphous catalysts is significantly different from crystalline catalysts, on a per site basis.

Finally, we compared the activity of the H-Y(CD) and US-Y samples for olefin oligomerization because this reaction is somewhat simpler. The measurements were made by holding the samples in a microbalance at a fixed temperature, exposing them to 10 Torr of either ethene or propene, then measuring the weight change for the samples after evacuation to determine whether oligomers had formed at the sites. For propene, oligomerization occurred rapidly at room temperature on both samples.[14] Oligomerization did not stop until a significant fraction of the pore volume of the materials, ~50%, had filled. Both materials were able to protonate propene, as shown in

Scheme 3, to form a carbenium ion, which could then react with other nearby propene molecules. The cracking of the oligomers in both samples was also indistinguishable. A wide range of products desorbed from the sample between 350 and 500K in TPD, similar to what is observed for H-ZSM-5.[16]   Of additional interest, neither faujasite sample was able to oligomerize ethene at 10 Torr, even at 400K.   Protonation of ethene requires the formation of a primary

Scheme 3

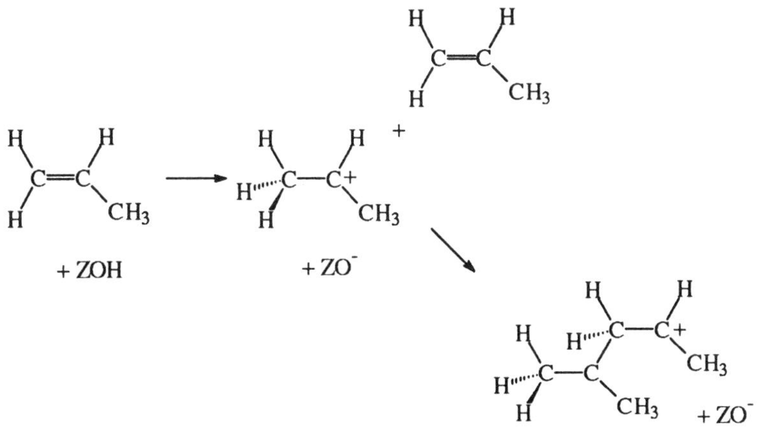

carbenium ion, rather than a secondary carbenium ion as in the case of propene.   Therefore, in the reactivity of propene and lack of reactivity in ethene, both H-Y(CD) and US-Y are indistinguishable in acidity.   This reinforces our earlier suggestion that increased rates for US-Y in alkane cracking are the result of something other than enhanced acidity of the Bronsted sites.

Finally, it should be acknowledged that Lewis sites can be catalytically active for some reactions.   For example, we have recently demonstrated that nonframework Al species in faujasites interact much more strongly with acetone than do the Bronsted sites.[17]   Based on the ^{13}C chemical shift of carbonyl-labeled acetone, there seems to be significantly more charge transfer on Lewis sites. One also observes the formation of condensation products much more rapidly when Lewis sites are present. Similar polarization effects could be important for the formation and stability of carbenium ions under the conditions of cracking reactions. However, a direct comparison is difficult to determine at the present time, but this is an interesting concept which deserves further attention.

V. ACKNOWLEDGMENTS

This work was sponsored by the NSF, Grant CTS94-03909

VI. REFERENCES

1. Abott, J.; Guerzoni, F.N.; *Appl. Catal. A,* 85 (1992) 173.
2. Juskelis, M.V.; Slanga, J.P.; Roberi, T.G.; and Peters, A.W.; *J. Catal.,* 138 (1992) 391.
3. R.J. Gorte, *Catalysis Today,* submitted for publication.
4. Farneth, W.E., and Gorte, R.J., *Chemical Reviews,* 95 (1995) 615.
5. Parrillo, D.J., Gorte, R.J., and Farneth, W.E., *J. Am. Chem. Soc.,* 115 (1993) 12441.
6. Parrillo, D.J., Lee, C, and Gorte, R.J., *Appl. Catal. A,* 110 (1994) 67.
7. Parrillo, D.J., Lee, C., Gorte, R.J., White, D., and Farneth, W.E., *J. Phys. Chem.,* 99 (1995) 8745.
8. Parrillo, D.J., Adamo, A.T., Kokotailo, G.T., and Gorte, R.J., *Appl. Catal.,* 67 (1990) 107.
9. Tittensor, J., Gorte, R.J., and Chapman, D., *J. Catal.,* 138 (1992) 714.
10. Biaglow, A.I., Gittleman, C., Gorte, R.J., and Madon, R.J., *J. Catal.,* 129 (1991) 88.
11. Biaglow, A.I., Parrillo, D.J., and Gorte, R.J., *J. Catal.,* 144 (1993) 193.
12. Periera, C., and Gorte, R.J., *Applied Catalysis A,* 90 (1992) 145.
13. Haag, W.O., and Chen, N.Y., in "Catalysis Design: Progress and Perspectives," I.L. Hegedus, ed., New York: Wiley, 1997, p. 81.
14. Biaglow, A.I., Parrillo, D.J., Kokotailo, G.T., and Gorte, R.J., *J. Catal.,* 148 (1994) 213.
15. Derouane, E.G., in "Guidelines for Mastering the Properties of Molecular Sieves," D. Barthomeuf, et al, eds., New York: Plenum, 1990, p. 234.
16. Gricus-Kofke, T.J., and Gorte, R.J., *J. Catal.,* 115 (1989) 233.
17. Biaglow, A.I., Gorte, R.J., and White, D., *J. Catal.,* 150 (1994) 221.

ATOMIC FORCE IMAGING OF A FLUID CRACKING CATALYST (FCC) SURFACE BEFORE AND AFTER AGING

M. L. Occelli(1), S. A. C. Gould(2), F. Baldiraghi(3), and S. Leoncini(3)

1. GTRI, Georgia Institute of Technology, Atlanta, Georgia 30332 USA
2. Keck Science Center, Claremont Colleges, Claremont, CA 91711, USA
3. EURON, San Donato Milanese, Milan, ITALY

I. ABSTRACT

An atomic force microscope (AFM) operating in a contact mode was used to observe changes in surface topography in FCC microspheres resulting from aging. AFM images of a commercial fluid cracking catalyst sample, before and after steam-aging, have been compared with those of the same FCC after use in a major refinery. Large scale images reveal that the FCC surface is formed by platelets and platelet aggregates and that the catalyst porosity result mainly from elongated and narrow cracks between these plates. Steam-aging does not seem to greatly change the FCC surface architecture. However, in the corresponding equilibrium catalyst (containing Fe, Ni, and V impurities) the FCC surface roughness decreased.

Atomic scale images were obtained for all the samples examined. The hexagonal and square symmetry observed in some images is consistent with the structure of Kaolin, a major component of these type of catalysts.

II. INTRODUCTION

Modern spectroscopic techniques are essential to the generation of detailed structural and compositional analysis necessary to the understanding and advancement of the science of catalyst design and catalysis. Today, scanning probe microscopy is becoming an ever increasingly important new tool to probe the surface architecture of catalysts at the macroscopic and atomic scale level. In fact, scanning tunneling microscopy (STM) has provided atomic resolution view of metal surfaces (1,2) while AFM has provided atomic scale details of nonconducting surfaces such as clays (3,4), zeolites (5-8), metal oxides (9), FCC (10,11), and hydrotreating catalysts (12). It is the purpose of this paper to report the use of AFM to observe the surface variations that a commercially available FCC sample undergoes after steam-aging at microactivity test (MAT) condition and after aging in a major European refinery.

III. EXPERIMENTAL

The AFM used for these experiments was a contact mode microscope (13,14) from Digital Instrument. FCC microspheres were glued onto steel disks with an epoxy resin. After the glue dried, the AFM tip was carefully placed in the middle of a microsphere. The probe was always placed on the very top of each FCC particle so that imaging occured by the probe and not by the edge of the cantilever. The images reported contain either 200 x 100 or 256 x 256 data points and nearly all images were acquired within seconds. The Si_3N_4 cantilevers (with integral tips) used for imaging were 120 μm in length and possessed a spring constant of 0.6 N/m. The force applied was in the 10 to 100 nN range. However, imaging was unstable unless the stiffest cantilever was used. More than 200 images have been examined in the present study. As before (4), the AFM was calibrated by imaging mica.

The FCC used in the present study, is a commercial sample currently in use in a major European refinery. Physical properties and composition data of the catalysts, before and after aging, have been presented elsewhere (16). Steaming was performed at 815°C by passing 100% steam at 1 atm for 5h over microspheres with size in the 100-200 mesh range.

IV. RESULTS AND DISCUSSION

Although informative and useful nitrogen porosimetry results provide little information on the topographical changes that could have occurred on the FCC surface during aging. This type of information can instead be obtained from atomic force microscopy images.

The images in Figures 1-5 were obtained with a contact mode AFM. When operating in a contact mode, the AFM cantilever is in constant contact with the sample surface where it fluctuates according to the surface topography and in a manner controlled by repulsive van der Waals's forces. This mode of operation provides high resolution but it can provide also artifacts resulting from morphological deformation induced by the rigidity of the cantilever in contact with the surface (17). Futhermore, it should be kept in mind that the AFM probably under estimate the height of the measured surface roughness because the AFM tip is pyramidal in shape with sides at 45°, and has a radius of curvature estimated at 20nm to 40nm (18). As a result, it may not be able to faithfully trace the steep and sometime deep grooves found on the FCC surface.

Thus, the FCC surface topography will be traced with an accuracy controlled by the tip size and by the cantilever spring constant. Rough FCC surfaces cannot, in general, be traced with accuracy because of the finite radius of curvature of the tip. The apparent depths of the cracks and valleys recorded in Figures 1-5 will depend on the size and rigidity of the cantilever used to trace the surface contours. In Figures 1-5 it is not always possible to determine if the observed surface depressions are pores, pits or valleys. The interpretation of these figures was assisted by taking side views and by rotating the images in different directions. By rotating the image, changing scan speeds and by

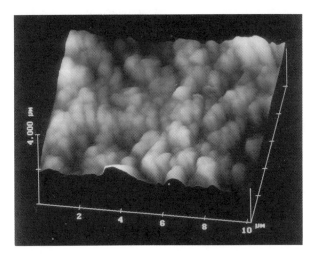

(A)

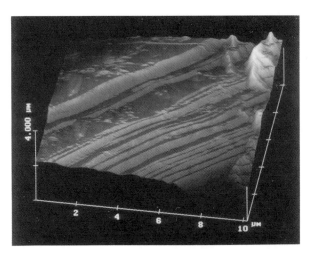

(B)

Figure 1.   AFM images of the surface of a Fluid Cracking Catalyst (FCC):
A) Surface roughness attributed to zones containing clay-gel-zeolite
mixtures, B) Surface detail resulting from Kaolin platelets, C) Long
range stacking of Kaolin platelets on the FCC surface, D) Surface
roughness from clay platelets aggregates, E) Valleys and pits
resulting from incomplete clay delamination, F) Plates can be of
different sizes and missing plates form pores on the surface.

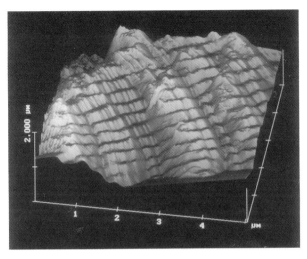

(C)

changing the imaging force, it is possible to distinguish between artifacts and real surface features.

The large scale AFM images in Figure 1 show the typical heterogeneity of the fresh FCC surface and that surface roughness is controlled mainly by the mode of clay (Kaolin) platelets aggregation. The distribution of these aggregates within the FCC microsphere seems to be responsible for the large cracks, pits, and valleys seen in these images, Figure 1. Rough domains as in Figure 1A, are attributed to zeolite-Kaolin-gel mixtures whereas the smooth surface seen in Figure 1B is due to a long range stacking face-to-face of kaolin platelets. These aggregates of clay platelets often times orient themselves as in

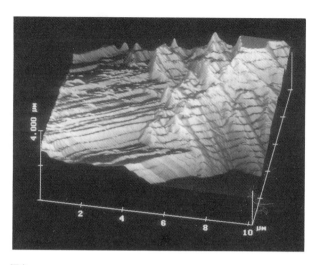

(D)

(E)

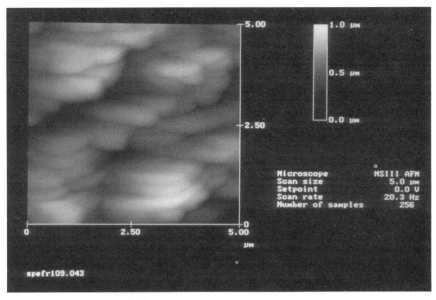

(F)

Figures 1C, 1D to drastically increase the surface roughness and can generate deep valleys and surface cracks as in Figure 1E. In Figure 1F it can be seen that clay platelets are of different sizes and that missing platelets form openings on the FCC surface.

Although pits, valleys and surface cracks are part of the FCC surface topography, slidths of the type represented in Figure 2 are the most common opening (or surface pores) on these type of surfaces. These slidths results from wedges (possibly gel particles) that props apart the clay plates as in Figure 2A

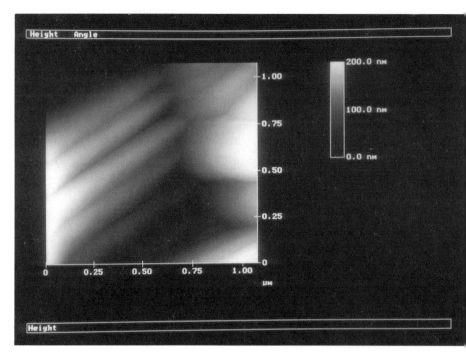

(A)

Figure 2.   AFM images of a fresh FCC surface showing: A) Voids between
            stacks of platelets are the source of opening to the catalysts internal
            structure, B) Elongated slits, C) Openings can result from the
            irregular agglomeration of clay platelets, D) Crater-like opening, E)
            Elongated cracks, the most common opening to access the catalyst
            internal porous structure, F) Atomic scale details of a plate.

or plate aggregates as in Figure 2B.   Valleys as in Figure 2C and craters as in

Figure 2D have been observed also in others type of FCC (10,11). However in

all the FCC microspheres observed, elongated slidths (as in Figure 2E) 1.0nm

(±0.2nm) in diameter irregular in size and shape, are the main source of porosity

in these catalysts. The atomic scale resolution image in Figure 2F, is void of

periodicity and it is believed to represent a clay plate covered with an

aluminosilicate gel (the FCC binder).

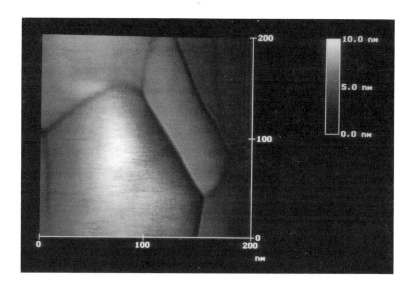

(B)

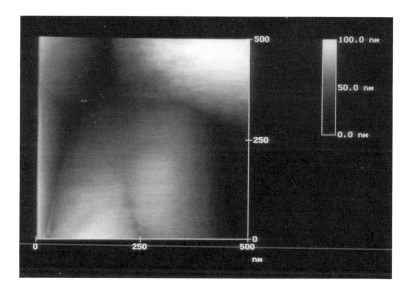

(C)

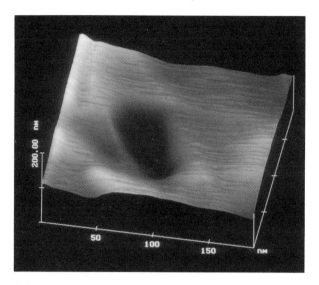

(D)

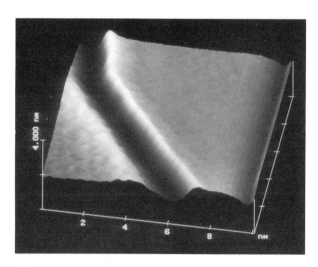

(E)

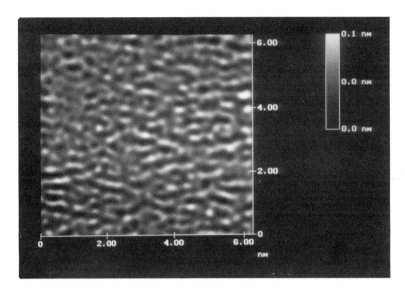

(F)

AFM images cannot readily differentiate FCC surfaces that have undergone a steam-aging treatment. The images in Figure 3 closely resemble those of the same FCC sample before steaming, see Figure 1. Thus it can be said that hydrothermal treatments at MAT conditions do not significantly alter the FCC surface topography.

The fresh and steam-aged FCC have a similar off-white coloration whereas the corresponding equilibrium sample have a gray color. In using an optical microscope to direct the tip of the cantilever on top of the FCC microspheres, it was observed that this gray coloration resulted from a mixture of white and totally black particles representing, we believe, the different residence times of these microspheres in the refinery fluid cracking unit (FCCU).

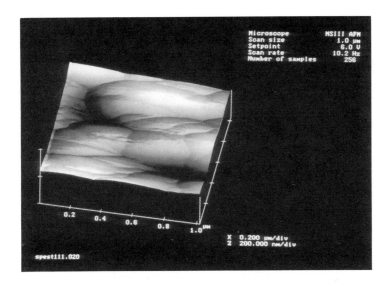

(A)

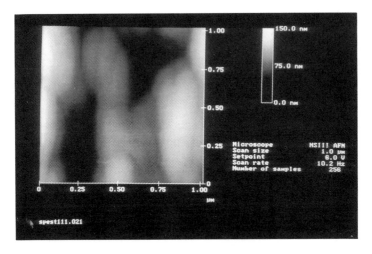

(B)

Figure 3.   Large scale AFM images of a fresh FCC after steam-aging: A) Side
view image showing pits and large cracks on the surface, B) The
top view image of A) and C) Large craters and cracks typical of
these type of surfaces.

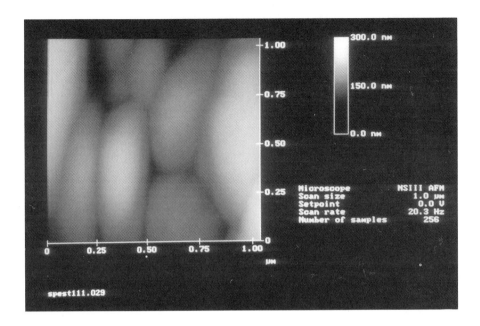

(C)

Coke and metals deposits on the FCC surface do not seem to alter the gross surface features in the equilibrium sample, Figures 4A, 4B.   Deep valleys and cracks are still visible, Figure 4C, 4D.   However, in contrast with fresh and steamed samples, the surface roughness is in general, decreased; see Figures 4E-4G. Smoother surfaces could be the result of coke, V, Fe, and Ni deposition (16).

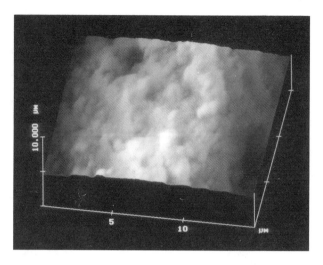

(A)

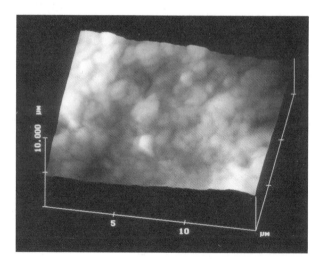

(B)

Figure 4.   Large scale AFM images of an equilibrium FCC from a major
             European refinery: A) Gross image of a white microsphere, B)
             Gross image of a black microsphere, C) Large craters and pits on
             the surface.   Surface roughness is still evident as in D, but in
             general surface roughness decreases as shown in E, F, and G.

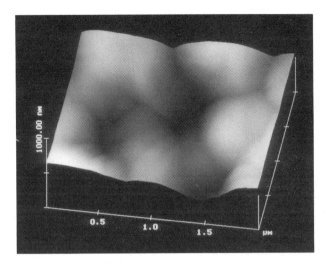

(C)

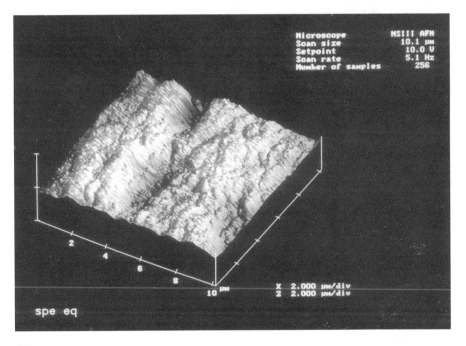

(D)

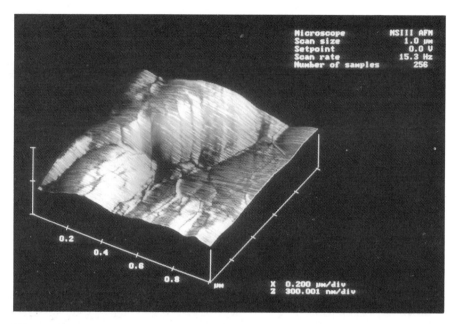

(E)

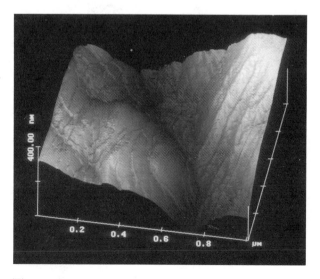

(F)

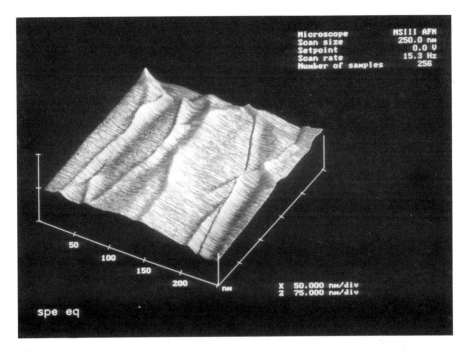

(G)

The atomic scale image in Figure 5A for white equilibrium samples, contains an hexagonal arrangement of white spots with nearest neighbor distance of about 0.5nm and lateral distance of 1.0nm representing the three basal oxygens of a SiO_4 unit in the kaolin silicate layer (3,4). The square arrangement of white spots in Figure 5B is believed to represent instead the oxygen's associated with the AlO_6 units in the kaolin octahedral layer (4,15). The disruption of the hexagonal pattern in Figure 5C and the equally observable disruption of the square pattern in Figure 5D have been attributed to coke and metals (such as Ni and V) deposition on kaolin plates in these black FCC microspheres.

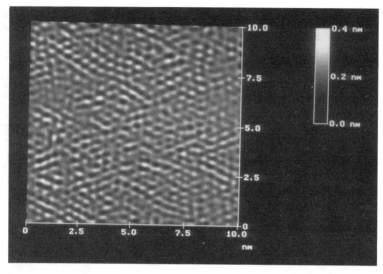

(A)

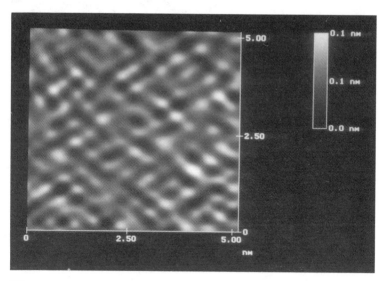

(B)

Figure 5.   Atomic scale AFM images of a white equilibrium FCC
             microsphere: A) The hexagonal arrangement of white spots, next-
             neighbor and lateral distances are consistent with the oxygen's of
             the SiO_4 layer in Kaolin, B) The square arrangement of white spots
             in the image are consistent with the oxygen's associated with the

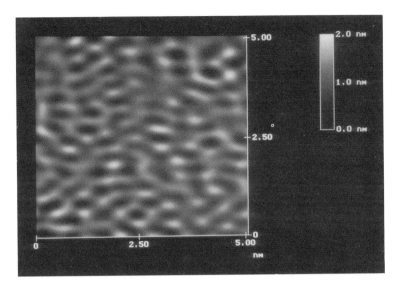

(C)

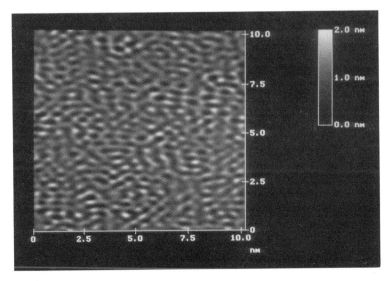

(D)

AlO$_6$ layer in Kaolin, C) In the black FCC microsphere, the SiO$_4$ layer is partially coated with coke. The hexagonal pattern is still visible but the repeat distances, (as seen in A) are no longer present, D) Distortion of the square pattern on the Kaolin octahedral layer.

V. SUMMARY AND CONCLUSIONS

It is believed that in FCC microspheres a mesoporous and macroporous structure with pores in the 2-10 nm range is responsible for hydrocarbons sorptions. Access to this structure is controlled mainly by elongated cracks 5 to 10 nm in width. Diffusion through these pores lead to the cracking centers located mainly in the micropores represented in gran part by the FCC zeolite content. The openness and the dimension of the pores that are available to sorption and catalysis will probably control the FCC ability to crack slurry oil and control LCGO yields. Acid site density and strength in the micropores is what determine gasoline generation.

Large (1-5 μm), crater like pores on the surface are believed capable to occlude and retain hydrocarbons that cannot be removed by steam stripping thus contributing to the FCC overall coke generation. These large pore have been attributed to clusters of Kaolin platelets imbedded on the FCC surface (10,11) and can be therefore avoided (or minimized) by a complete delamination of the clay during catalysts preparation.

AFM images from this work have shown that the surface topography of an FCC does not suffer major variations upon steaming. Large scale AFM images indicate that surface roughness decreases somewhat after steaming and that smoother surfaces are more easily observed in the corresponding equilibrium FCC samples. Atomic scale AFM images have allowed the identification of Kaolin, a major component of this type of catalysts.

The major effect of aging at MAT conditions and of aging in a refinery. FCCU is the drastic reduction of the fresh FCC acid sites number, strength and distribution (16) together with a reduction and modification of the catalyst internal pore structure.

VI. REFERENCES

1. Weimer, A. J., Kramer, C. Bai, J. D. Baldeschwieler, and W. J. Kaiser in "Proc. 2nd Int. Conf. on STM," R. M. Feenstra Ed., p. 336 (1988)
2. Stupian, G. W., and M. S. Leung in "Proc. 2nd Int. Conf. on STM," R. M. Feenstra Ed., p. 371 (1988)
3. Hartman, H., G. Sposito, A. Yang, S. Manne, S. A. C. Gould, and P. K. Hansma, Clays and Clay Minerals, 38, 4, 337 (1990)
4. Occelli, M. L., B. Drake, and S. A. C. Gould, J. Catal, 142, 337 (1993)
5. Komiyama, M. and Tashima, T. Jpn. J. Appl. Phys., Vol. 22, pp. 3761-3763, (1994)
6. Weisenhorn, A. L., MacDougall, J. E., Gould, S. A. C., Cox, S. D., Wise, W. S., Massie, J., Maivald, P., Elings, V. B., Stucky, G. D., and Hansma, P. K., Science 247, 1330, (1990)
7. MacDougall, J. E., Cox, S. D., Stucky. G. D., Weisenhorn, A. L., Hansma, P. K., and Wise, W. S., Zeolites, Vol. 11., June (1991)
8. Scandella, L., Kruse, N., and Prins, R., Surface Science Letters 281, L331-L334, (199)
9. Carmi, Y., Dahm, A. J., Eppell, S. J., Jennings, W., Marchant, R. E., and Michael, G. M., J. Vac. Technol., B10, 2302 (1992)
10. Occelli, M. L. and Gould, S. A. C., CHEMTECH, May 24 (1994)
11. Occelli, M. L. and Gould, S. A. C. in ACS Symposium Science, Vol. 571, M. L. Occelli and P. O'Connor Eds., p. 271 (1994)
12. Chianelli, R., in "Proc. Int. Symp. Adv. Hydrotreating," (in press)
13. Meyer, G. and Amer. N. M. (1988), Appl. Phys. Lett. 53, 1095 (1988)
14. Alexander, et al., J. Appl. Phys., 65, 164-167.
15. Occelli, M. L. and S. A. C. Gould (unpub. results)
16. Occelli, M. L., A. Auroux, F. Baldiraghi, and S. Leoncini, this volume (1996)
17. Gould, S. A. C., Schiraldi, D., and Occelli, M. L., in preparation.
18. Akamine, S., Barrett, R. C., and Quate, C. F., Appl. Phys. Lett., 57, 316. (1990)

MODIFIED USY ZEOLITES FOR FCC CATALYSTS BY ACID EXTRACTION

Bing-Lan Li, Xing-Zhong Xu, Jianming Su, and Huifang Pan

Research Institute of QiLu Petrochemical Corp.
Zibo, Shandong, 255400, People's Republic of China

ABSTRACT

The ultrastable Y (USY) Zeolite is always accompanied with nonframework aluminium (NFAL) species in its micropores. The chemical extraction of Al atoms from NFAL and framework aluminium (FAL) using aqueous inorganic acid solution was studied. After hydrothermal treatment at 800°C for 17 hours, the microactivity of FCC catalyst comprising the acid modified USY zeolite as the active component can be increased by 2–3 units and the specific coke formation can be reduced by about 40% compared to that of the catalyst containing USY without acid treatment.

The dealumination mechanism of acid attack on USY zeolite was presented by a CNDO/2 quantum chemistry calculation, which suggests that the dealumination process from framework can be performed by an acid attack on the negatively charged oxygen sites bonded to the Al atoms.

INTRODUCTION

USY zeolite is the main active component of FCC catalysts. Now it is common to blend 10%–25% of residue oil into the gas oil feed of the FCC unit, which results in a higher level of coke deposited on the catalyst. The decreasing coke formation of zeolite has received considerable attention recently, especially for the cracking of heavy oil. It is known that FCC catalysts commonly comprise a USY zeolite and an inorganic oxide matrix. We discovered that a catalytic cracking process using a catalyst comprising a zeolite pretreated properly with an inorganic acid produces less coke than that utilizing a catalyst of the same zeolite but without acid treatment. The acidic hydroxyl groups of zeolites are the centers of coking and the coke deposit occurs mainly on the strong acid sites of the catalyst in the hydrocarbon cracking process. The reduction of the strong acid sites by means of dealumination of Y zeolite to increase the framework Si/Al molar ratio can decrease the coking rate. Therefore the coking rate during the cracking reaction is an inverse measure of Si/Al ratio for the same crystalline structure zeolite Y (1).

USY zeolite has a higher Si/Al ratio than Y zeolite. During the dealumination stage of hydrothermal ultrastabilization of zeolite Y, the nonframework aluminium (NFAL) species, such as Al_2O_3 $[Al(OH)_2]^+$, $[Al(OH)]^{2+}$, and $(AlO)^+$ are formed and occupy the micropore volume of USY zeolite, which have been shown by Al magic angle spinning nuclear magnetic resonance (MASNMR) experiments (2,3). Here, the chemical extraction of Al atoms from NFAL and framework aluminium (FAL) of USY zeolite using aqueous inorganic acid solution was experimentally and theoretically studied to reveal the effect of enhancing Si/Al ratio on the coking rate and activity of n-heptane cracking. The extent of dealumination was found to be related to the concentration and the addition rate of the acid solution. The optimum acid treatment conditions of USY zeolite have been examined with an aim to enhance Si/Al ratio, retain the crystallinity, decrease coking rate of the n-heptane cracking reaction, and increase the cracking activity of USY zeolite. Therefore, the FCC catalyst using the acid modified USY zeolite as an active component has greatly succeeded in improving the FCC catalyst. A reaction mechanism of the acid attack was presented by CNDO/2 quantum chemistry calculations (4).

EXPERIMENTAL

USY zeolite was prepared by hydrothermal ultrastabilization of NH_4Y, which was obtained from NaY after ion exchange with NH_4^+ for 90%, at a certain temperature for a few hours. Each USY zeolite sample was then stirred in an aqueous nitric acid solution to extract aluminum. The acid extracted USY zeolites were filtered, washed, and dried to form DUSY zeolites. The elemental compositions of the samples were determined by chemical analysis and the contents of NFAL and FAL can be calculated from the chemical compositions of zeolites. The crystallinity of each sample was examined by XRD. The unit cell dimension was estimated by XRD using the Bragg equation from the diffraction angle of a (555) crystal plane. The framework Si/Al ratio can be calculated from unit cell dimension. The coking rates of dealuminated USY (DUSY) zeolites with various framework Si/Al ratios during the cracking of n-heptane were determined by the method of thermogravimetry. The measurements of the n-hexane cracking activity measurements of DUSY zeolites were performed in a fixed bed microreactor. The first order rate constant k of each sample was obtained from the total conversion of n-hexane cracking at the equilibrium stage (5). The FCC catalysts comprising 35 wt% of different DUSY zeolites and kaolin clay matrix were prepared. Experimental catalysts were aged by hydrothermal treatment in 100% steam at 800°C for 4 and 17 hours. The microactivity (MA) and the coke formation of the aged catalysts were tested according to the standard MAT procedure.

RESULTS AND DISCUSSION

The Effect of the Addition Rate of Acid on the Chemical Extraction of Al Atoms from NFAL of USY Zeolite

When USY zeolite is treated with 0.3N nitric acid solution of lower concentration the NFAL content decreases because a number of Al atoms are extracted from NFAL species as shown in Table 1. In comparison, the samples 0, 1, 2, 3, and 4 have similar unit cell size and framework Si/Al ratio; however, the NFAL content of DUSY zeolite decreases with the increase of the addition rate of acid solution. The reason for this is that NFAL species occupying the micropore volume of zeolite are preferentially removed by the reaction with acid solution as follows:

$$Al_2O_3 \quad + 6HNO_3 \rightarrow 2Al^{3+} + 6NHO_3^- + 3H_2O$$
$$[Al(OH)]^{2+} + NHO_3 \rightarrow Al^{3+} + NO_3^- + H_2O$$
$$[Al(OH)_2]^+ + 2NHO_3 \rightarrow Al^{3+} + 2NO_3^- + 2H_2O$$
$$(AlO)^+ \quad + 2NHO_3 \rightarrow Al^{3+} + 2NO_3^- + H_2O$$

Table 1. Physical and Chemical Properties of DUSY Zeolites with Addition Rate of 0.3N HNO₃

Sample	Adding rate of 0.3N HNO₃ (m l/g.h)	Unit cell size (nm)	Framework SiO₂/Al₂O₃	Crystallinity (%)	NFAL content (wt%)
USY,0	Untreated	2.4401	8.73	68.1	13.68
DUSY,1	0.87	2.4403	8.60	73.6	8.16
DUSY,2	1.25	2.4406	8.27	73.5	5.25
DUSY,3	2.50	2.4402	8.66	73.5	4.87
DUSY,4	4.44	2.4405	8.46	74.9	4.34
DUSY,5	added in whole-wise	2.4358	13.09	48.4	—

The integrity of zeolite is retained after dilute acid extraction, and the crystallinity of DUSY zeolites slightly increases with the removal of NFAL species. When USY zeolite is treated with HNO_3 of the same concentration but the acid is added in whole-wise (DUSY 5), a portion of the framework is collapsed because a number of the framework Al atoms are extracted by the violent acid attack that leads to the lower crystallinity of DUSY zeolite.

The Effect of Acid Concentration on the Chemical Extraction of Al Atoms from NFAL and FAL of USY Zeolite

The effect of different acid concentrations on the chemical and physical properties including the coking rate and the cracking activity of n-hexane for DUSY zeolites has been investigated as shown in Table 2.

Table 2. Physicochemical and Catalytic Cracking Properties of DUSY Zeolites

Sample	HNO_3 concentration (N)	Adding speed of HNO_3 (ml/g.h)	Unit cell size (nm)	Framework (SiO_2/Al_2O_3)
USY, 0	0.0	Untreated with acid	2.4401	8.73
DUSY, A	0.1	2.61	2.4400	8.80
DUSY, B	0.3	2.50	2.4402	8.66
DUSY, C	0.6	2.45	2.4401	8.73
DUSY, D	1.0	2.37	2.4318	23.15
DUSY, E	2.0	2.27	2.4247	45.91
DUSY, F	4.0	2.37	2.4242	60.07

Sample	Crystallinity (%)	NFAL content (wt%)	Surface area (m^2/g)	Micropore volume (cm^2/g)	Coking rate (mg/min.g)	Rate constant (ml/min.g)
USY, 0	68.1	13.68	594.1	0.27	0.95	12.9
DUSY, A	73.6	8.16	636.8	0.30	0.88	13.2
DUSY, B	73.5	4.87	675.0	0.32	0.52	13.8
DUSY, C	74.0	2.21	708.5	0.34	0.45	12.8
DUSY, D	73.3	0.82	795.1	0.38	0.34	11.8
DUSY, E	55.4	Partial framework collapses	—	—	0.12	10.3
DUSY, F	50.8	Partial framework collapses	—	—	0.09	8.5

These data show that under the condition of about the same addition rate of acid solution, samples O, A, B, and C have similar unit cell sizes and framework Si/Al ratios with HNO_3 concentrations below 0.6N, due to the fact that when the HNO_3 concentration is below 0.6N, only the NFAL species are extracted from the pores of zeolites and H^+ ions could not attack the zeolite framework. The apparent decrease of unit cell size and increase of framework Si/Al of sample D treated with 1N HNO_3 indicate that not only NFAL species are extracted by also a few of the framework Al atoms are removed, while the integrity of USY zeolite can still be retained by a slight attack of H^+ on the framework of zeolite. But when excess H^+ ions (HNO_3 concentration 2.0N) is used to treat USY zeolite, the partial framework collapses leading the samples DUSY-E and DUSY-F to lower crystallinities and lower cracking rate constants.

When HFAL plugging the zeolite pores is removed by acid extraction, the micropore volumes of DUSY zeolites increase in the order of the sample D>C>B>A>O. A more open network of pores can improve the diffusion of large hydrocarbon molecules and cause the decrease of coke yield, so the cracking rate constants of sample O through B increase. However, the cracking rate of samples C and D are lower than that of samples O through B, because the zeolite NFAL is also a kind of active alumina and can promote the catalytic cracking of hydrocarbon. When the NFAL of zeolite is greatly extracted by higher H^+ concentration treatment, the large hydrocarbon molecules are difficult to convert and the overall catalytic cracking efficiency declines (6). Comprehensively considering the activity and the anticoking behavior, the optimum HNO_3 concentration is in the range of 0.3–0.4N.

The Catalytic Properties of FCC Catalysts Comprising DUSY Zeolites

The FCC catalysts comprising acid modified DUSY zeolite (35 wt%), kaolin clay, and binding material were prepared. The industrial USY (crystallinity = 71%, unit cell size = 2.4462nm) were used as the starting material for the acid extraction to form DUSY. The prepared catalysts were aged by hydrothermal treatment in 100% steam at 800°C for 4 and 17 hours. The catalytic properties of the catalysts are shown in Table 3 and compared with the catalyst using unmodified USY as an active component.

Comparing the properties of Cat-1 and Cat-2 with Cat-0 in Table 3, it is clear that the microactivities of FCC catalysts containing modified USY zeolites treated by 0.3–0.5N HNO_3 can be increased by 5–6 units (4 h aged) and 2–3 units (17 h aged), respectively and the specific coke can be reduced by 30–40%.

Table 3. Catalytic Properties of FCC Catalysts Comprising DUSY Zeolites

Sample	Zeolite	HNO₃ concentration (N)	MA 800°C /4h	MA 800°C /17h	Specific coke* 800°C /4h	Specific coke* 800°C /17h
Cat-0	USY	0.0	57.3	45.0	1.19	1.43
Cat-1	DUSY	0.3	63.8	47.3	0.846	0.845
Cat-2	DUSY	0.4	62.4	48.2	0.849	0.768
Cat-3	DUSY	0.5	60.4	41.8	0.910	0.742
Cat-4	DUSY	0.6	55.2	45.7	1.221	1.483

*Specific coke = $C\% \times 100/MA$.

The excellent catalytic cracking performances of FCC catalysts made of DUSY zeolites indicate that the proper extent of acid treatment of USY zeolite is a very effective chemical modification method for the production of FCC catalyst with high activity and coke selectivity.

The Quantum Chemistry Calculations of Acid Dealumination in USY Zeolite

The quantum chemistry calculations may give some specific quantitative information for the above dealumination of USY zeolite by acid treatment. The CNDO/2 calculations were made on a model of the six-ring cluster with a $T_6O_6(OH)_{12}$ structure unit (T represents an Al atom or Si atom) of Figure 1 (4,7). This model is used in the calculations to get charge densities and bond orders of Al-O and Si-O in the six-ring cluster. Only the S,P basis orbitals are taken into account in all quantum chemistry calculations. The charge densities of each skeletal atom of the cluster are listed in Table 4. The oxygen atoms are considered to be the reaction sites for H^+ attack dealumination because of the negative charges on them. Beran (7) suggested that hydroxyl groups are formed on the position of oxygen atoms bonded to Al. When H^+ is introduced into zeolite skeleton with Si/Al=5, it attracts oxygen (O_2) of the cluster as shown in Figure 1. When an H^+ is added on the reaction site, most of the bond orders of T-O are reduced compared to those of the cluster without the H^+ (Table 5). The bond order of the O_2-Al bond in particular becomes about one-half of the original. The O_2-Al bond weakening becomes more significant. The reduced bond order facilitates the Al-O bond cleavages.

The enhanced negative charge density (q_4) on an O_4 atom after breaking an Al-O bond (Table 6) indicates that the most probable site in the second

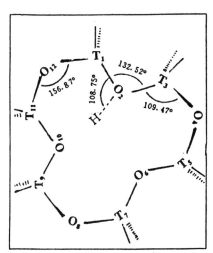

Fig 1.Schematic depiction of six-ring cluster model
for Si/Al=5(T3 is an Al atom)

Table 4. CNDO/2 Charge Densities (q) for the Cluster Mode $T_6O_6(OH)_{12}$ with Si/Al=5

q1	q2	q3	q4	q5	q6	q7	q8	q9	q10	q11
1.5580		1.3594		1.5462		1.6283		1.6396		1.6322
	0.7327		−0.6836		0.7076		−0.7324		—	
									0.7248	

Table 5. CNDO/2 Wiberg Bond Orders (p) for the Cluster, $T_6O_6(OH)_{12}$ and H^+-$T_6O_6(OH)_{12}$ with Si/Al=5

Cluster	P_{1-2}	P_{2-3}	P_{3-4}	P_{4-5}	P_{5-6}	P_{6-7}
$T_6O_6(OH)_{12}$	1.0828	0.5878	0.5878	1.0528	0.7300	0.9107
H^+-$T_6O_6(OH)_{12}$	0.5813	0.3288	0.6970	0.9862	0.7407	0.8712

Cluster	P_{7-8}	P_{9-10}	P_{10-11}	P_{11-12}	P_{12-13}	P_{1-12}
$T_6O_6(OH)_{12}$	0.8077	0.86600	0.8555	0.8109	0.9134	0.7331
H^+-$T_6O_6(OH)_{12}$	0.7948	0.8813	0.7705	0.5478	0.7665	0.9157

Table 6. CHDO/2 Charge Densities (q) for Cluster Model with Si/Al=5 After Breaking of an Al-O Bond

q1	q2	q3	q4	q5	q6	q7	q8	q9	q10	q11	q12
1.6720		1.2871		1.5999		1.6496		1.6562		1.6586	
											-0.7240
	0.5239		0.7639		0.6781		0.7228		0.7510		

acid attack is O_4, since the O_4-Al bond is weakened. Due to the second acid attack, the O_4-Al bond is broken, then the aluminium atom (T_3) is disconnected from the zeolite skeleton.

The acid attack mechanism deduced from the quantum chemistry calculations can be used to theoretically demonstrate the effect of acid treatment on chemical and physical characterization of USY zeolite. When USY zeolite is treated with HNO_3 of lower concentration and lower addition rate, only the NFAL species in the pores are extracted and H^+ ions could not attack the oxygen sites of the zeolite skeleton. The crystalline integrity of zeolite can be retained. When using 1N HNO_3 for the acid treatment of USY zeolite (Table 2), most of the NFAL species are extracted and only a few of the FAL atoms are removed by a few of the H^+ ions attacking the framework of zeolite; therefore the zeolite framework could not be destroyed. Slightly increasing the crystallinity and apparently increasing the framework Si/Al ratio of USY zeolite are very effective on enhancing the anticoking property of USY zeolite (1) (Table 2). Adding excess H^+ ions (NHO_3 concentration 2.0N) to USY zeolite, the partial crystalline framework of zeolite collapses under the violent simultaneous attack of excess H^+ ion on the oxygen atoms of the framework, which induces the reduction of the crystallinity and cracking activity of the DUSY zeolite.

The concentration and addition rate of acid solution should be properly controlled during the acid extraction of USY zeolite. The selected optimum conditions of acid treatment for USY zeolite have been used to prepare FCC catalysts, which exhibit excellent catalytic cracking performances.

CONCLUSIONS

1. The proper extent of inorganic acid treatment of USY zeolite is an effective chemical modification method for the improvement of the catalytic cracking performances of FCC catalysts.

2. When USY zeolite is treated with a nitric acid solution of lower concentration and lower addition rate, only the nonframework aluminum species are extracted and H^+ could not attack the zeolite skeleton. The nonframework aluminium content of USY zeolite decreases with the increasing of the concentration of nitric acid solution. Partial framework may be collapsed by the excess H^+ ions violent simultaneous attack on the oxygen atoms of the framework, which leads to the lower crystallinity and cracking activity of USY zeolite.

3. The optimum acid treatment conditions should be properly controlled to retain the crystallinity, enhance the Si/Al ratio, decrease the coking rate, and increase the cracking activity of UCY zeolite.

4. The CNDO/2 quantum chemistry calculations suggest that acid attacks are performed at the oxygen sites that are attached to Al atoms of the framework. The theoretical calculation results are achieved in assuming the acid attack mechanism.

REFERENCES

1. Pan Huifang, Su Jianming, and Wang Biao, *Acta Petro, Sinica 8*(3), 29 (1992).
2. Addison, S.W., Cartliedge, S., Harding, D.A., McElhiney, G., *Appl. Catal., 45*, 307 (1988).
3. Corma, A., Fornes, V., Rey, F., *Appl. Catal., 59*, 267 (1990).
4. Jong, T.K., Myung, C.K., Yasuaki, O., T.O. Shino, *Bv. I. J. Catal., 115*, 319 (1989).
5. David, K., Edward, F.T., Lee, and Lovat, V. C. Rees, *Zeolites, 8*(3), 288 (1988).
6. Willis, W.S., and Suib, S. L., *J. Am. Chem. Soc., 108*, 5657 (1986).
7. Beran, S., *J. Phys. Chem. Solids, 43*, 221 (1982).

ORIGIN AND CONTROL OF NO_x IN THE FCCU REGENERATOR

A. W. Peters
G. Yaluris
G. D. Weatherbee
X. Zhao

Grace Davison, Columbia, MD 21044

ABSTRACT

In this paper we describe some of the chemistry in the FCCU regenerator leading to NO_x formation. This chemistry provides a technical basis for various NO_x control strategies in the FCCU including additives. NO_x is typically formed in concentrations of 100 to 500 ppm in the flue gas from the regenerator. We have completed the nitrogen balance around an experimental pilot size FCCU. Almost half of the nitrogen is found in the light and heavy cycle oils and the other half is found as coke on the spent catalyst. Very little of the NO_x emitted comes from oxidation of air nitrogen, probably much less than 30 ppm. We have determined that as much as 90% of the nitrogen in the coke is emitted as N_2. The rest is reduced to nitrogen in the regenerator. Most of the NO_x is formed in the regenerator as NO, with very little N_2O or NO_2 being formed initially. Other reduced nitrogen species such as cyanides or amines may be formed in small quantities as transient species. CO, coke on catalyst, and other hydrocarbon species can be effective for NO_x removal in the presence of the proper catalyst. NO_x reduction may also involve the selective conversion of intermediate nitrogen species to N_2 rather than NO. Effective strategies for reducing NO_x emissions from an FCCU regenerator include additives that catalyze pathways leading to the formation of N_2.

INTRODUCTION

Both nitrogen and sulfur oxides are environmental pollutants subject to regulation. Commercial additives exist to control SO_x emissions from the FCCU regenerator to a low level. NO_x emissions are typically lower than uncontrolled SO_x emissions and have received less attention. However, NO_x participates more strongly in ozone formation and in that sense is an important pollutant. NO_x levels in the regenerator flue gas from the FCCU are typically in the range of 100 ppm to 500 ppm. NO_x includes both NO and NO_2. NO_2 is formed only after the NO is emitted to the atmosphere, and N_2O is found only at very low levels in the FCCU regenerator. Consequently, NO_x control and control of NO in the regenerator are in practice the same thing.

An important source of NO is the nitrogen in the FCCU feed. The chemistry of NO formation from the feed has recently been discussed in two independent studies (1,2). The results reported in these studies are in essential agreement. In order to assess the importance of feed nitrogen it is necessary to answer the questions, how much nitrogen goes to products, how much appears in the regenerator as coke, and how much of the nitrogen in the coke is converted to NO. Nitrogen is present in FCCU feeds in amounts ranging from about 0.05% to as much as 0.35%. Most of this nitrogen occurs in aromatic structures and is expected to show up in either the bottoms fraction or in the coke on the catalyst. In the regenerator coke is combusted to CO_2 and small amounts of CO, depending on the promotion level and the level of excess oxygen. Just as sulfur in coke is oxidized in the regenerator to SO_2 and SO_3, the nitrogen in coke is oxidized to NO.

The fate of feed nitrogen in the FCCU may depend on a number of factors like catalyst properties, the chemical nature of the nitrogen species in the feed (pyrolic versus pyridinic), unit design and operation conditions. However, nitrogen mass balance studies in a pilot plant circulating riser around both the riser side and the regenerator side provide important information on what happens to nitrogen during the cracking process and the subsequent catalyst regeneration. Results reported earlier (1) show that while nearly half of feed nitrogen appears in the coke on the catalyst, only a small portion of the coke nitrogen is converted to NO in the flue gas. The rest is converted to N_2. There are several possible reaction pathways through which nitrogen in the coke can react. These reaction pathways may be similar to reaction schemes reported for coal combustion (3) and can lead to N_2 and/or NO formation. In order to understand the pathways for the formation of these products, we have performed experiments that include careful nitrogen mass balances around the catalytic cracking and catalyst regeneration processes, as well as Temperature-Programmed Oxidation (TPO) experiments of spent catalysts in conventional fixed bed and novel fluidized bed apparatuses. We have also performed thermodynamic calculations and experiments to determine if "thermal" NO_x is formed in the FCCU regenerator. Thermal NO is NO that results from direct oxidation of air nitrogen to NO in some types of high temperature combustion (e.g., coal and natural gas combustion for energy generation, and glass ovens).

Understanding the formation of NO in the regenerator has important implications on the development of effective strategies to control NO emissions. For example, managing the amount of reductant species (coke, CO, etc.) in the regenerator is very important in achieving a decrease of NO emissions (4). One strategy to reduce NO_x emissions is the use of catalyst additives. However, the use of catalytic additives to control other emissions, may affect the level of reductants in the regenerator (e.g., conventional CO combustion promoters for decreasing CO emissions). Increases of NO emissions are observed when Pt/alumina promoters are used to reduce CO emissions. Therefore, catalytic additives designed to reduce NO$_x$ emissions must be effective in the presence of additives used to control other pollutants that have the side effect of increasing NO$_x$ emissions. In this paper we examine how the underlining chemistry of NO formation and destruction may be affected by the presence of various additives in the regenerator, and what are the catalytic properties a NO$_x$ reduction additive must have to be effective.

NITROGEN MASS BALANCES IN THE FCCU

Nitrogen mass balances were obtained on the two different feed stocks described in Table 1. The feed stocks were cracked in the Davison Circulating Riser (DCR), a pilot unit scale FCCU (5). This unit processes about 1 Kg of feed per hour. The riser and regenerator temperatures were 521 °C (970 °F) and 732 °C (1350 °F) respectively. One of the feed stocks was a relatively high nitrogen gas oil (0.32% N) and the other a gas oil with an average nitrogen content (0.131% N). The nitrogen content of the feeds and liquid products was determined at Huffman Laboratories by a modified Kjeldahl N method (ASTM D3179 Method B) for pyrolic type nitrogen species, and amine titration with a solution of perchloric acid in acetic acid for pyridinic type nitrogen species. Measurements of the nitrogen in the coke on catalysts were done by a LECO Carbon-Nitrogen Analyzer. Product yields are presented in Table 2 along with nitrogen mass balance data. For both feeds only about 45% of the feed nitrogen appears in the total liquid product. Although there is some concentration of nitrogen in the bottoms compared to the feed, most of the hydrocarbon molecules left in the bottoms fraction are smaller than the original species. These differences in molecular weight between the bottoms and the feed imply that the percentage of nitrogen containing molecules in the feed is about the same as in the bottoms.

Nitrogen can also appear as ammonia in the light gases or as nitrogen in coke. The ammonia in the light gases and the stripping gas was measured by bubbling the gas product after separation through a 0.1 N HCl water solution and determining the amount of nitrogen in the liquid. The results of this procedure from an experiment in the DCR using the moderate nitrogen level feed at a 73% conversion level showed that about 5% of the feed nitrogen is released during steam stripping, and about 3% appears as ammonia as a result of cracking.

These results suggest that about 50% of the feed nitrogen should be recovered as coke. Since typical NO$_x$ emissions amount to only about 5% or less of the feed nitrogen, this result implies that most of the nitrogen in the coke appears in some other form, possibly N$_2$. Normally this amount of N$_2$ is not measurable since the flue gas may contain *ca.* 70 wt. % N$_2$ from the air used in

Table 1:   Feedstock properties.

Feedstock:	Moderate N	High N
API @60 F	25.8	21.1
Aniline Pt. F	189	164
Sulfur wt. %	0.301	0.774
Total Nitrogen wt. %	0.13	0.32
Basic Nitrogen	0.052	0.13
Conradson Carbon wt. %	0.53	0.12
'K' Factor	11.78	11.32
Ni ppm	0.88	<1
V ppm	0.53	<1
Fe ppm	0.58	5
Na ppm	3.6	<1
Simulated (GC) Distillation		
Vol. %, Temp °F		
10	568	550
50	790	757
90	986	923

the regenerator to burn the coke. A direct measurement of low levels of nitrogen on the coked catalyst is difficult. With about 50% of the feed nitrogen contained in the 5% fraction of the feed converted to coke, nitrogen is concentrated in the coke by a factor of about 10. However, with only *ca.* 1% coke on catalyst, even this degree of concentration only represents about 100 to 200 ppm N on the catalyst, an amount difficult to accurately measure.

We measured the formation of N_2 during catalyst regeneration by conducting a set of nitrogen mass balances with a mixture of 5% O_2 in Ar used to regenerate the catalyst during the operation of the DCR. The results of these experiments are shown in Table 3. Molecular nitrogen was excluded from the DCR regenerator. The appearance of significant amounts of nitrogen in the flue gas measured by gas chromatography confirmed the formation of N_2 from coke in the regenerator. In both balances 30 wt. % to 40 wt. % of the feed nitrogen was recovered as N_2 and achieved balance closure of 80% to 90%. The measurement of N_2 in the flue gas agreed well with the direct determination of nitrogen on the catalyst in the case of the feed with moderate nitrogen concentration, but in the case of the feed with high nitrogen the measured nitrogen on the spent catalyst was lower than expected. These results suggest

Table 2: Nitrogen mass balance experiments, and nitrogen distribution in the products from the FCCU.

Feedstock:	Moderate Nitrogen	High Nitrogen
Catalytic Cracking Conditions		
C/O	6.8	7.0
Wt. % Conversion	66.5	57.4
H2	0.08	0.18
C1-C4	13.9	12.2
C5+ Gasoline	49.05	40.3
% of feed N in Gasoline	**1.5**	**2.0**
LCO	17.9	22.3
% of feed N in LCO	**6.3**	**12.0**
Bottoms	15.6	20.3
% of Feed N in Bottoms	**26.7**	**27.5**
Total % Feed N in Distilled Product	**34.5**	**41.0**
Undistilled Liquid Product*		
% of Feed N	**43.0**	**47.4**
Coke , Wt. % of Feed	3.26	4.1
% of Feed N in Coke	**36**	**41**
Regenerator Gas		
NO, ppm	**100**	**190**
% of Feed N	**4**	**3**

*Separately measured from undistilled products

that as much as 90% of the nitrogen in the coke is converted to N$_2$ in the regenerator.

The chemistry of nitrogen in the FCCU regenerator is different from the chemistry of sulfur (6). Fifty to sixty percent of the sulfur in the feed appears in the liquid products, and most of the rest, 35% to 45% appears as H$_2$S. Only 2% to 5% appears in the coke, and nearly all of the sulfur in the coke is oxidized to sulfur oxides. The case of nitrogen is very different. However, somewhat less nitrogen is present in the liquid products, very little appears as ammonia, and nearly half is present in the coke. Surprisingly, very little of the nitrogen in the coke is oxidized to NO. Most is converted to N$_2$ in the regenerator.

Since half of the feed nitrogen appears in the products as coke and since NO is formed from the coke nitrogen, one could expect that NO emissions

Table 3: Nitrogen mass balance experiments in the DCR using oxygen/argon in place of air.

Feedstock: **Catalytic Cracking Conditions**	**Moderate Nitrogen**	**High Nitrogen**
C/O	8.0	7.7
Wt. % Conversion	73.9	59.8
H2	0.1	0.12
C1-C4	18.1	13.2
% of feed N in Light Gases		**3.0**
C5+ Gasoline	51.67	40.2
% of feed N in Gasoline		**3.6**
LCO	15.5	22.6
% of feed N in LCO		**12.0**
Bottoms	10.6	19.4
% of Feed N in Bottoms		**31.5**
		50.1
Total liquid product*		
% of Feed N	**47.4**	-
Coke , Wt. % of Feed	3.68	4.26
% of Feed N in Coke	**35**	-
Regenerator Gas	**Ar/O$_2$**	**Ar/O$_2$**
Flue Gas NO (ppm),	62	211
% of Feed N recovered as **NO**	**2.6**	**2.7**
Flue Gas N$_2$ (%)	0.045	0.125
% of Feed N recovered as N$_2$	**38**	**32.7**
Total N Recovery, Wt. %	**88.0**	**85.6**

*Separately measured from undistilled products

should correlate with the nitrogen content of the feed. However, because different feeds contain different kinds of nitrogen and different relative amounts of pyrolic and pyridinic nitrogen, the amount of nitrogen in the feed may only loosely correlate with the nitrogen in the coke, and further, the amount of NO formed relative to N$_2$ may depend on the amount and properties of the coke on the catalyst and on conditions in the regenerator. The observed relatively loose

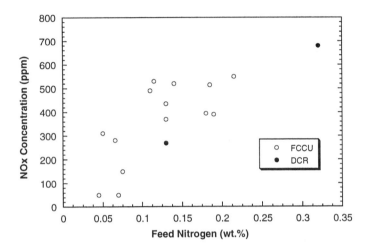

Figure 1:   Correlation of NO emissions with feed nitrogen. CO combustion
promoted regenerator. FCCU data is from reference 7.

correlation is consistent with these observations. The results shown in Figure 1
are from both commercial FCCU operations (7) and from the Grace Davison
DCR.

FORMATION OF THERMAL NO IN THE REGENERATOR

One possible source of NO in the regenerator is the oxidation of the
nitrogen in the air to produce NO. The results of thermodynamic calculations
for this process are shown in Figure 2. Even if this reaction goes completely to
equilibrium, less than 10 ppm NO will be produced under regenerator
conditions of 1% excess O$_2$ and temperatures between 730 °C (1340 °F) and
780 °C (1430 °F). Even at 870 °C (1600 °F) less than 30 ppm of NO is
expected from this reaction.   Since the concentration of NO in typical
regenerators is considerably greater than the maximum equilibrium value of 30
ppm, the effect of this reaction, if it occurs, must be to convert NO back into
molecular nitrogen (N$_2$).   Consequently, the oxidation of air makes no
contribution to the observed NO$_x$ in the FCCU flue gas.

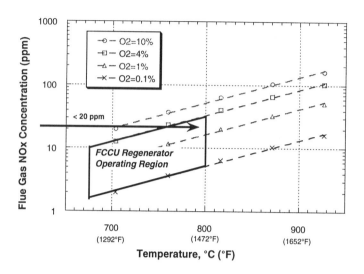

Figure 2: Calculated equilibrium concentrations of NO for 75% N_2 and various levels of excess regenerator O_2.

THE COMPETITIVE FORMATION OF N_2 AND NO DURING CATALYST REGENERATION

The nitrogen mass balance results described, show that N_2 is produced in the regenerator from the nitrogen in the coke during coke combustion. This surprising result can be explained by several reaction schemes. One such reaction scheme is that during catalyst regeneration, and in the presence of coke, nitrogen can be removed as either N_2 or as NO. Another possible pathway is that nitrogen is removed from the catalyst during regeneration as NO and it is subsequently reduced to N_2 by reaction with carbon on unregenerated catalyst, CO, or other reductant species. A third pathway is the formation of reduced nitrogen containing compounds in the vapor phase of the regenerator which subsequently are either oxidized to NO or finally form N_2. It is clear that NO and N_2 are the two stable products of coke burning in the FCCU regenerator, and that the final level of NO will depend on the competition between the various pathways leading to N_2 and NO as the final products.

The process of regeneration, burning coke from the catalyst in the presence of oxygen, can be followed in the laboratory by a TPO experiment. We conducted TPO experiments in a conventional fixed bed reactor apparatus equipped with a mass spectrometer. The amount of catalyst loaded in the fixed bed reactor of the TPO system was 0.6 g. The spent catalyst was regenerated by heating the catalyst at a heating rate of 30 °C/min in a flow of 30 sccm of 5% oxygen in helium. As the temperature increases, the amount of CO, CO_2, and NO formed during the reaction is monitored by the mass spectrometer. The results of this TPO experiment (Figure 3) show that carbon burns to CO_2 at a temperature about 50 °C (90 °F) less than nitrogen burns to NO. The NO observed during the experiment amounts to about 1% of the total carbon on the catalyst on a mole basis and agrees reasonably well with the direct determination of nitrogen on the catalyst. In this experiment nearly all of the nitrogen in the coke appears to have been burned to NO. N_2 cannot be directly observed in the presence of CO, because both have the same molecular weight, 28 AMU.

In another TPO experiment, Figure 4, the catalyst is heated to 550 °C (1010 °F) in the presence of the oxygen helium mixture and held at that temperature for 20 minutes. At this point the temperature is again increased. During the initial heating most of the carbon is burned to CO_2, and very little NO is formed. During the second stage most of the NO is formed and relatively little CO_2. These results show that carbon and nitrogen are burned sequentially, first carbon, then nitrogen. It appears from these experiments that the oxidation of nitrogen to NO occurs at a higher temperature than the oxidation of carbon to CO_2 or to CO. These results would imply that as the amount of carbon on the regenerated catalyst decreases, the nitrogen content of the remaining coke increases. Measurements of the nitrogen in coke by a LECO analyzer have shown that this is the case (Figure 5). Nitrogen is removed last during regeneration.

During the fixed bed TPO experiments, we were unable to observe nitrogen species other than NO due in part to the small sample size and the very small amount of material to be detected. Potential products like NH_3 and HCN tend to adsorb on the walls of the experimental system making their detection difficult. We have recently developed a novel fluidized bed reactor equipped with an IR detector that avoids this difficulty (8). This apparatus was developed in an effort to simulate FCC catalyst regeneration in the laboratory, and gain a better understanding of the chemistry operative in the regenerator. Thus, it was not constructed specifically as a TPO apparatus. The presence of a cyclone to separate entrained particles form the reactor gas effluent can distort the peaks' position (shift to higher temperatures) and shape (tailing effects). However, the order in which the various species appear, and the peaks are detected, remains unaffected. Therefore, when this apparatus is used in TPO experiments properly, it can provide important information about the chemistry of coked catalyst regeneration.

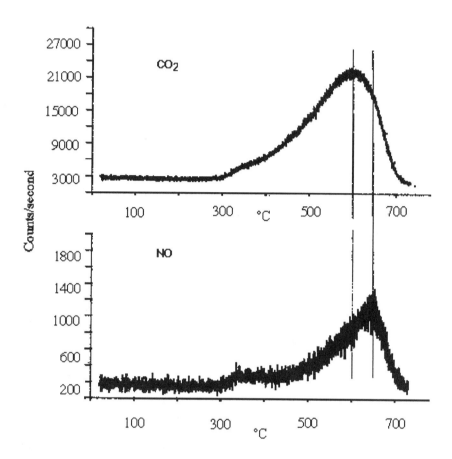

Figure 3:   Temperature-Programmed Oxidation in flowing O_2/He of a spent
FCC catalyst in a fixed bed apparatus.

In these TPO experiments about 20g of fluidized spent (coked) catalyst
were fluidized continuously using 800 sccm of *ca.* 5% O_2 in N_2. The heating
rate was 9 °C/min until the reactor temperature reached 780 °C (1435 °F). The
reactor temperature was then held constant at 780 °C. The FTIR active gases in
the reactor effluent were analyzed by an On-Line 2002 Multigas analyzer. The
fluidized bed conditions of these experiments are closer to the actual conditions
during catalyst regeneration in an FCCU. Furthermore, in this system enough

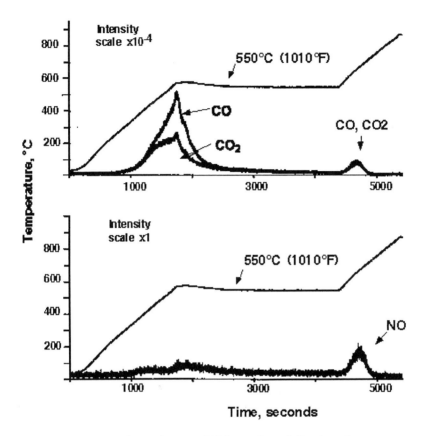

Figure 4: Regeneration of spent FCC catalyst following the temperature profile shown.

flue gas is produced during regeneration to observe the formation of both N_2O and HCN at relatively low temperatures. Figure 6 shows that, in agreement with the fixed bed TPO results, nitrogen species are observed at a higher temperature than CO and CO_2 (about 50 °C). In addition, the amount of nitrogen detected (area under curve in Figure 6) which does not include any molecular nitrogen formed, is about 0.75% on a mole basis of the amount of carbon species detected. However, the results in Figure 7 show that contrary to the fixed bed TPO experiments, NO is not the only nitrogen species observed. The first nitrogen species observed peaks at 670 °C, and is HCN, which is also the most abundant of the nitrogen species detected (about 80% of the total FTIR active nitrogen species detected). N_2O is detected next, peaks at 705 °C, and is about

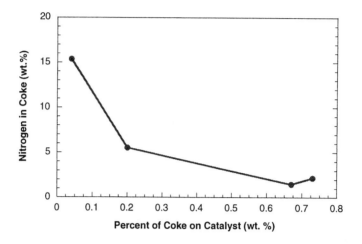

Figure 5:   Nitrogen in coke remaining after partial regeneration of the catalyst.

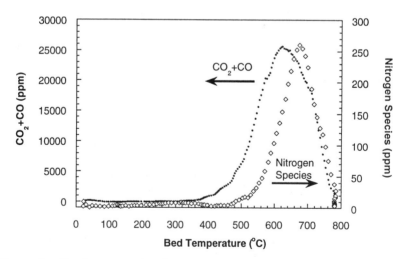

Figure 6:   Total amount of carbon oxides and nitrogen species detected during Temperature-Programmed Oxidation of FCC coked catalyst in a fluidized bed apparatus.

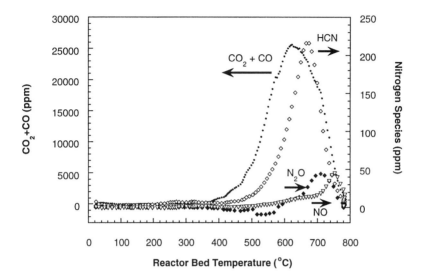

Figure 7: Total amount of carbon oxides, HCN, N$_2$O, and NO nitrogen species detected during Temperature-Programmed Oxidation of FCC coked catalyst in a fluidized bed apparatus.

10% of the total nitrogen species. NO is detected last at higher temperatures when most of coke on the catalyst has been burned off. It peaks at 750 °C and is about 10% of the total nitrogen species.

The temperature at which HCN is detected is relatively low compared to typical regenerator conditions. During steady state coked catalyst regeneration experiments at 700 - 715 °C (1292 - 1319 °F), we have detected small amounts of both HCN and N$_2$O in the reactor effluent. Therefore, while it is unclear if HCN is actually present in the regenerator in significant amounts, the data presented here show that HCN can be one of the intermediates participating in the regenerator chemistry.

NO REDUCTION TO N$_2$ IN THE REGENERATOR

During the laboratory regeneration simulations, the coke oxidizes to NO only after most of the carbon is already burned to CO or CO$_2$. Since most of the nitrogen in the coke ends up as N$_2$ rather than NO, there must be a subsequent

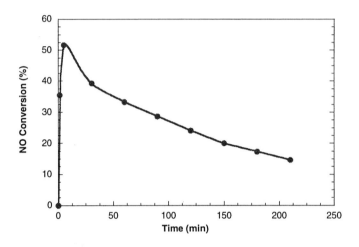

Figure 8:   Conversion of 1951 ppm NO flowing at 440 cc/min over 5.0 grams of spent FCC catalyst at 732°C (1350°F).   The catalyst before reaction contained 0.573 wt.% carbon and 0.211 wt.% after the reaction.

reduction step via which at least some of the nitrogen in the coke is directly oxidized to NO. Either gas phase NO is reduced to N_2 via reaction with the coke on the catalyst, or the nitrogen while still on the catalyst reacts directly with coke to form N_2, or both. In the case of the TPO experiments, the second possibility may be less important since the carbon on the catalyst is nearly gone before the observed oxidation of nitrogen to NO.   However, in an FCCU regenerator there may be sufficient amounts of coke present to make this reaction pathway important. The reaction of NO with carbon has been reported previously as being as rapid or perhaps more rapid than the reaction of coke with oxygen (9). In an effort to directly observe the reaction of carbon (coke) with NO, a stream of NO was reacted with carbon on a spent catalyst at 732 °C (1350 °F) over a period of time. The results of this experiment are shown in Figure 8. During the experiment about 2.2 mmol of NO reacted with 1.5 mmol of C. This implies a mixed stoichiometry for the formation of approximately equal amounts of CO and CO_2 from the reactions:

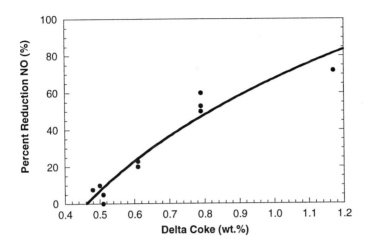

Figure 9:   A DCR correlation between NO reduction and coke as a weight percent of the catalyst (delta coke) for a variety of FCC catalysts.

$$2\,C + 2\,NO \;\rightarrow\; 2\,CO + N_2$$
$$C + 2\,NO \;\rightarrow\; CO_2 + N_2$$

NO conversion decreases with time as the amount of coke on catalyst decreases.

In the operating DCR regenerator this result translates into an operating relationship between NO reduction and delta coke or the amount of coke on the catalyst (Figure 9). NO decreases as delta coke increases. The patent literature describes a number of NO$_x$ reduction strategies, including regenerator design changes, that exploit this chemistry (10-13). Some are based on placing more reductant in the regenerator, and in other cases the regenerator flue gas is contacted with a reductant in the form of CO or spent catalyst.

CONTROL OF NO EMISSIONS

Effect of Excess Oxygen on NO Emissions

One option that can be used to increase the amount of coke and other reductant species in the regenerator is to reduce excess oxygen or to operate in a partial burn. The effect of excess oxygen on NO concentration in the DCR is shown in Figure 10. For the DCR regenerator, it is necessary to operate at

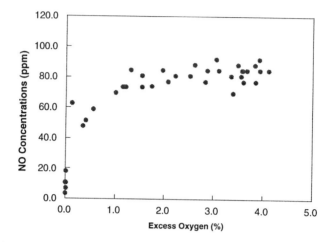

Figure 10: The effect of excess oxygen on NO emissions in the DCR.

nearly zero excess oxygen to significantly influence the NO concentration. Operation at or below 0.5% excess O_2 only achieves about a 20% reduction compared to operation at 1% - 2%.

The effect of operation in partial burn is illustrated in Table 4. The NO level drops from 85 ppm to 20 ppm as the regenerator changes from full to partial burn operation (*ca.* 2% CO) using the moderate nitrogen feed stock of Table 1. Differences in activity and in selectivity are the result of differences in the carbon levels on regenerated catalyst.

The Effect of Combustion Promoter on NO Emissions

The effect of CO combustion promoter on NO flue gas levels in the DCR is dramatic (Figure 11). Using the moderate nitrogen feed stock (Table 1) NO levels increase by a factor of four with the addition of 0.5 wt. % conventional Pt-based promoter. The effect of CO promoter is not entirely understood. The CO promoter decreases the concentration of reducing species (coke, CO and others) in the dense bed of the regenerator. Thus, it removes some of the reactants necessary for NO reduction to N_2 after it is formed. The fluidized bed TPO experiments show that reduced nitrogen species (e.g., HCN) may act as intermediates in some NO formation reactions. In this case the CO combustion

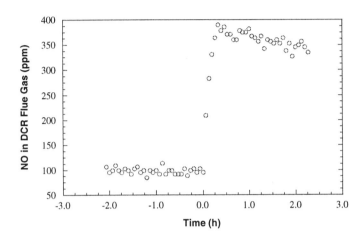

Figure 11: Effect of conventional Pt-based CO combustion promoter (0.5 wt.% additive level) on DCR NO$_x$ emissions.

promoter may participate by oxidizing these reduced species directly to NO rather than N$_2$.

Catalytic Additives for Reduction of NO Emissions

To reduce NO emissions when a CO combustion promoter is used in the FCCU regenerator, it is necessary to develop catalytic additives that can promote NO reduction by the two mechanisms described. They must be active and selective for NO reduction by CO and other reducing species, even under relatively oxidizing conditions and when the amount of these reducing species is lowered dramatically by the use of CO combustion promoters. Alternatively, and to the extent NO is formed over combustion promoters by oxidation of reduced nitrogen intermediates like HCN, they must be able to remove these intermediates before they react, by converting them to N$_2$ rather than NO.

Such additives are now commercially available and new additives are under development. An active additive developed for its ability to selectively reduce NO to N$_2$ in the presence of CO when added to a catalyst inventory, can significantly decrease NO formation. Such catalytic additive is shown in Figure

Table 4:   DCR NO and CO emissions in full and partial burn. (Moderate nitrogen feed, 970°F reactor, 1300°F regenerator, 300°F feed)

	Partial Burn	Full Burn
% Yields		
Conversion	69.2	74.2
C/O ratio	6.5	6.7
H2	0.14	0.15
C1-C4	15.8	17.5
C5 Gasoline	49.1	51.4
LCO	17.1	15.2
Coke % of Feed*	4.07	5.00
Regenerator		
% CO_2	6.03	8.08
% Excess O_2	0	1.44
% CO	1.93	0.02
NO	20 ppm	85 ppm

* Note that in partial burn, the coke as a percent of the feed is given by measured levels of CO plus CO_2 in the flue gas of the DCR, while in full burn only CO_2 is used.

12. In the experiment shown, a Grace Davison proprietary additive was added at 0.5% level to the catalyst inventory already containing 0.5% of a 500 ppm Pt containing CO combustion promoter. The results show, that during pilot plant testing in the DCR, and at the same level of CO oxidation activity provided by the combustion promoter, the NO reduction additive achieves a decrease of NO emissions of about 45%.

CONCLUSIONS

The question of the origin of the NO in the FCCU regenerator flue gas has been answered by performing carefully designed nitrogen mass balances around a pilot plant unit. The nitrogen mass balance results show that after processing in the FCCU most of the nitrogen in the gasoil is found either in the liquid products, especially the cycle oil, or is carried into the regenerator as coke. In the experiments described here, the feed nitrogen is split evenly, about 35 wt. % to 50 Wt. % in both the liquid products and in the coke. Little or none of the NO is formed by oxidation of N_2 in the air because temperatures in the regenerator are not high enough. The evidence is that molecular nitrogen is produced in the regenerator either from a secondary reduction of NO with coke and other reducing species (e.g., CO), or directly from the nitrogen in the coke.

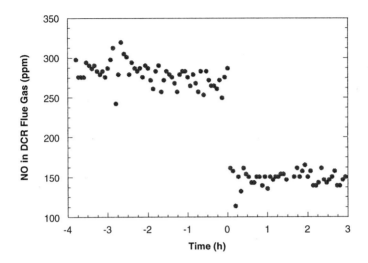

Figure 12: The effect of a proprietary commercial additive at 0.5 wt.% additive
level on NO$_x$ emissions from the DCR regenerator.

In the absence of any catalytic additives, the former reaction pathway appears to
be more important.   Nearly all (up to 90%) of the nitrogen in the coke is
ultimately reduced to molecular nitrogen, and only a very small amount appears
as the pollutant NO. HCN, and other reduced nitrogen species can be formed in
the regenerator. Their role in the overall chemistry is unclear, but they may play
a role in NO formation when CO combustion promoters are used.   Thus,
selective conversion by catalytic additives of these intermediates to N$_2$, rather
than NO, can be one mechanism by which the additive decreases NO emissions.
     The results imply that NO can be controlled by controlling the amount of
total reducing species available in the regenerator, a result previously disclosed
in the patent literature (10-13). Conditions favoring lower NO, include reduced
excess   oxygen,   including   partial   burn   conditions,   and   improved   CO
management policies utilizing controlled additions of a CO promoter. Excess
promoter will minimize the amount of CO or other reductants available in the
regenerator and may oxidize intermediate nitrogen containing compounds
directly to NO. The effect of CO promoter can be mitigated by the addition of
Grace Davison proprietary catalytic additives. These additives reduce the NO in
the flue gas of the circulating bed riser (DCR) by over 40%.

ACKNOWLEDGMENTS

We would like to thank M. Juskelis, J. Swaine, K. Smith, K. Kreipl, and J. Hawkins for their contributions to this work.

REFERENCES

1. Peters, A.W., Weatherbee, G.D., and Zhao, X., *Fuel Reformulation* **5(3),** 45 (1995).

2. Tamhankar, S., Menon, R., Chou, T., Ramachandran, R., Hull, R., and Watson, R., *Oil & Gas J.,* March 4, 1996, p. 60.

3. Wojtowicz, M.A., Pels, J.R., and Moulijn, J.A., *Fuel Processing Technology* **34,** 1 (1993).

4. Evans, R.E., and Quinn, G.P. *in* "Fluid Catalytic Cracking: Science and Technology" (Magee, J.S. and Mitchell, M.M., Jr. Eds.), *Stud. Surf. Sci. Catal.* Vol. 76, p. 563. Elsevier Science Publishers B. V., Amsterdam, 1993.

5. Young, G.W. *in* "Fluid Catalytic Cracking: Science and Technology" (Magee, J.S. and Mitchell, M.M., Jr. Eds.), *Stud. Surf. Sci. Catal.* Vol. 76, p. 257. Elsevier Science Publishers B. V., Amsterdam, 1993.

6. Wormsbecher, R.F., Weatherbee, G.D., Kim, G., and Dougan, T.J., "Emerging Technology for the Reduction of Sulfur in FCC Fuels," NPRA, AM-93-55, 1993.

7. Bernstein, G., "FCCU Data," US EPA, Dockett # A-79-09, 1982.

8. Yaluris, G., Peters, A.W., and Zhao, X. *in* "Preprints, Impact of Clean Air Act on Fuels Production and Use," 212th ACS National Meeting, Orlando, FL, August 25 - 29, 1996, Vol. 41(3), p. 901.

9. Chu, X., and Schmidt, L.D., *Ind. Eng. Chem. Res.* **32,** 1359 (1993).

10. An example is a counter-current regenerator design by Kellogg claimed to reduce NO_x by 50% or more. Spent catalyst is distributed over the top of the regenerator bed. NO reacts with the coke on the spent catalyst.

11. Buchanan, J.S., and Johnson, D.L., US 5 372 706 (1994).

12. Terry, P.H., US 5 360 598 (1994).

13. Owen, H., and Schipper, P.H., US 5 077 252 (1991).

EFFECT OF STRIPPER CONDITIONS ON THE YIELD AND STRUCTURE OF COKE DERIVED FROM N-HEXADECANE

Abdul Aziz H. Mohammed [1], Brian J. McGhee [1], John M. Andresen [1], Colin E. Snape [1] and Ron Hughes [2]

[1] University of Strathclyde, Dept. of Pure & Applied Chemistry, Glasgow G1 1XL, UK
[2] Chemical Engineering Unit, University of Salford, Salford, M5 4WT, UK

INTRODUCTION

Coked catalyst in a fluid catalytic cracking (FCC) unit first passes to a steam stripper to remove residual volatiles and then it is transferred to the regenerator vessel where the coke is burned in a stream of air. Since the catalyst acts as a heat-transfer medium with the heat liberated by coke combustion providing the energy for the endothermic cracking reactions, the coke selectivity can markedly affect an FCC unit's profitability. It is now generally accepted that, as well as being formed via the actual cracking reactions associated with the strongly acidic catalytic sites, coke can arise from (i) normal thermal reactions, (ii) dehydrogenation reactions promoted by metals - particularly Ni in heavy feeds and (iii) entrained products which are symptomatic of incomplete stripping and can contribute to the overall level of coke (O'Connor and Pouwels, 1994). The entrained products increase the hydrogen content of the coke and the additional air requirement gives rise to excessively high temperatures in the regenerator and additional steam, which, in turn contributes significantly to the deactivation of FCC catalysts.

Although the deactivation of FCC catalysts via coke deposition has been the subject of much investigation since the 1940s (Butt, 1972; Wolf and Alfani, 1982), there is still a lack of knowledge on the contributions of the mechanisms outlined above to the overall level of coke formation. This situation has arisen from the inherent difficulties of characterising the structure of insoluble cokes at the low concentrations encountered in FCC. Indeed, to facilitate coke characterisation, most fundamental studies thus far on the ultra-stable (US) type Y zeolites have involved excessively high levels of carbon deposition in relation to normal FCC operation (Groten *et al*, 1990). Moreover, the behaviour with small molecules, where coke can be formed directly from the reactant within the zeolite framework, is different to that observed with heavy feedstocks where the yield of coke deposited on the catalyst (typically *ca* . 1%) is independent of catalyst/oil ratio (Turlier *et al*, 1994) and the conversion is governed primarily by the extra-framework mesopores formed by steaming USY zeolites (Dai *et al*, 1994).

To determine how the yields and composition coke varies in FCC as a function of stripper conditions, tests have been conducted in fluidised-bed reactor using n-hexadecane at 520°C, a commercial FCC catalyst with stripping periods of up to 120 min. Coke concentrates have been prepared by demineralisation with hydrofluoric and hydrochloric acids for characterisation by ^{13}C NMR using the inherently quantitative single pulse excitation (SPE) or Bloch decay technique (Love *et al*, 1993), with complementary information being obtained from probe mass spectrometry. This approach was successfully demonstrated recently for FCC refinery catalysts (Snape *et al*, 1995, McGhee *et al*, 1995) where the cokes were found to be highly aromatic in character (carbon aromaticities >0.95) . However, differences in feedstock composition were still reflected in the structure of the cokes with the aromatic nuclei being more highly condensed in coke from a vacuum residue than that from a hydrogenated vacuum gas oil. This investigation on coke derived from n-hexadecane extends our earlier studies on the roles of quinoline, phenanthrene and other model compounds as poisons and coke inducers for n-hexadecane (Hughes *et al*, 1994[a], 1994[b] and 1995).

EXPERIMENTAL

Stripping Tests and Catalyst Demineralisation
The commercial zeolite US-Y catalyst used for the fluidised-bed tests on n-hexadecane to simulate stripping (Crosfield's CBZ 2) contained 29% alumina matrix. The all-quartz reactor had a 4 cm diameter bed and, during each run at 520°C, *ca* 15.5 g n-hexadecane was fed into

the reactor containing 80 g of catalyst over a period of 10 min. The total flow of nitrogen through the bed was 4 dm^3 min^{-1} with a flow of 1 dm^3 min^{-1} being used to introduce with the liquid feed into the centre of the bed. The catalyst was then kept at the reaction temperature under the nitrogen flow for the periods up to 120 min. to simulate stripping. To investigate how the composition of the coke might be affected by catalyst formulation, a test was also carried out with a second commercial FCC catalyst (pseudo boehmite alumina matrix, 20%), but with a total flow of nitrogen through the bed of 6 dm^3 min^{-1}.

The coked catalysts were first refluxed in chloroform for 3 hours to remove any trapped molecular species and then vacuum-dried prior to demineralisation, their carbon contents being determined before and after chloroform extraction. The standard demineralisation procedure for solid fuels (Saxby, 1976, Durand *et al*, 1980) was then applied to the chloroform-extracted catalysts. This involved:
(i) extraction with 2M hydrochloric acid (stirring overnight at 60°C) and;
(ii) extraction of the hydrochloric acid-treated sample with 40% hydrofluoric acid (HF) at room temperature for 4 hours with 20 cm^3 of HF being used per gram of sample.
The coke concentrates were finally washed with dilute hydrochloric acid to remove any remaining inorganic paramagnetics prior to collection in plastic filtration equipment. The vacuum-dried coke concentrates recovered from the catalysts had carbon contents in the range of 30-50%.

^{13}C NMR

Cross polarisation (CP) and SPE ^{13}C NMR measurements on the coke concentrates were carried out as previously described (Love *et al*, 1993; Snape *et al*, 1995) at 25 MHz using a Bruker MSL100 spectrometer with magic-angle spinning (MAS) at 4.5-5.0 kHz to give spectra in which the sideband intensities are only *ca* 3% of the central aromatic bands. Approximately *ca* 150 mg of sample was packed into the zirconia rotors. The ^1H decoupling and spin-lock field was *ca* 60 kHz and, for SPE, the 90° ^{13}C pulse width was 3.4 ms. A recycle delay of 50 s was employed between successive 90° pulses in SPE to ensure that virtually complete thermal relaxation occurred (Snape *et al*, 1995). Normal CP spectra of the coke concentrates were obtained using contact times of either 1 or 5 ms.

Dipolar dephasing (DD) was combined with both SPE and CP to estimate the proportion of protonated and non-protonated aromatic carbon using a dephasing period of 50 μs. No background signal was evident in the SPE spectra from the Kel-F rotor caps The measurement of aromatic and aliphatic peak areas manually was found to be generally more precise than using the integrals generated by the spectrometer software.

Mass Spectrometry
Mass spectrometry was conducted on the coke concentrates using a VG instrument in which the probe was heated from ambient to 500°C at a rate of 20°C min^{-1} and spectra over the mass range 50-600 were recorded every 5s. Spectra were recorded in both electron impact (EI) and chemical ionisation (CI, with ammonia) modes.

RESULTS AND DISCUSSION

Carbon Balance
The overall conversion of n-hexadecane was 75% w/w under the experimental conditions used prior to stripping, this being similar to that obtained previously in the MAT reactor at the same reaction temperature of 520°C (Hughes *et al*, 1994[a], 1994[b] and 1995). Table 1 lists the mean carbon contents from duplicate determinations on the initial coked catalysts at the two flow rates used and the catalysts recovered after stripping for 60 and 120 min at the lower flow rate. The value of ca 1.5% for the initial coked catalyst at the lower flow rate corresponds to a coke yield from n-hexadecane of ca 8% w/w. There is a decrease of 0.25% after 60 min., but the carbon content then remains fairly constant close to 1.25%. Chloroform extraction did not affect the carbon contents by more than the experimental error of $ca \pm 0.05\%$ obtained for different samples from the same run recovered from the reactor (Table 1), indicating that, unlike for heavy petroleum feeds (Turlier *et al*, 1994), entrained products are not present in significant amounts.

The second catalyst used gave a considerably higher selectivity for a similar conversion (71%) w/w, the carbon content of 2.5 % corresponding to a coke selectivity of 12%.

Table 1 Carbon contents of coked catalysts obtained from n-
hexadecane

Catalyst	Flow rate/ dm^3 min^{-1}	% Cw/w
Cat. 1, initial	4	1.54
Cat. 1, stripped for 60 min.	4	1.31
Cat. 1, stripped for 120 min.	4	1.26
Cat. 2, initial,	6	2.53

* = mean of determinations on two different samples recovered from
the reactor for each run.

Initial Cokes
Figure 1 shows the CP spectrum of the coke concentrate obtained
from the first catalyst and and Figure 2 compares the CP and SPE
^{13}C NMR spectra of the coke concentrate obtained from the second
one used. These cokes are much more aliphatic character than for
the refinery cokes analysed previously (Snape $et al$, 1995) with the
aromaticity (total sp^2 carbon) values derived from the CP and SPE
spectra both being close to 0.82 (Table 2). However, the sp^2 carbon
envelope for the second coke concentrate contains a small
contribution from carboxyl and carbonyl groups (Figure 2, ~ 2-3
mole % carbon, 175-205 ppm). This has probably arisen from
accidental exposure of the coked catalyst to hot air during recovery
from the fluidised-bed reactor. Assuming that aliphatic groups have
oxidised preferentially, this means that the aromaticity is probably
somewhat lower (ca 0.80) than that of the lower flow rate coke.

The presence of a relatively high proportion of aliphatic carbon in the
coke concentrates is consistent with the EI mass spectrometry results
(Figure 3) where the alkyl fragments (m/z 41, 43, 57, 71 and 85) are
the most intense fragments as found for the corresponding MAT
cokes analysed previously (Hughes $et al$, 1994[a], 1994[b] and
1995). Even at high probe temperatures, the mass spectra are still
dominated by fragments characteristic of alkylated species with only
minor peaks being observed for >2 ring polynuclear species (e.g.
141, 155, 178, 202, 226; in the bottom trace of Figure 2, peaks
beyond m/z 130 have been expanded 100 fold).

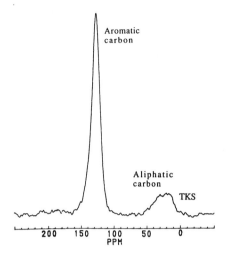

Figure 1 CP ^{13}C NMR spectrum (1ms contact time) of the coke concentrate from the first catalyst with a nitrogen flow rate of 4 dm^3 min^{-1}.

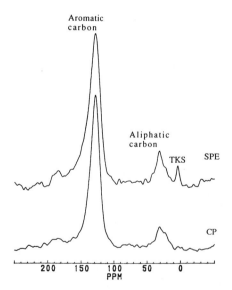

Figure 2 CP (1ms contact time) and SPE ^{13}C NMR spectra of the coke concentrate obtained from the second catalyst with a nitrogen flow rate of 6 dm^3 min^{-1}.

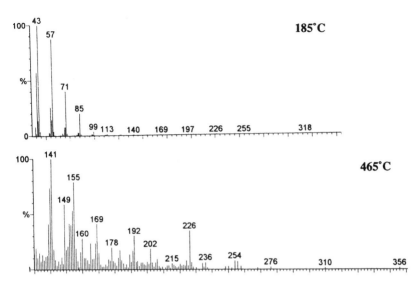

Figure 3 EI mass spectra of the coke concentrate obtained from the second catalyst. The top scan shown was taken at 185°C. In the bottom scan taken at 465°C, the peaks arising from ≥ 2 ring aromatics have been expanded by approximately 100 fold.

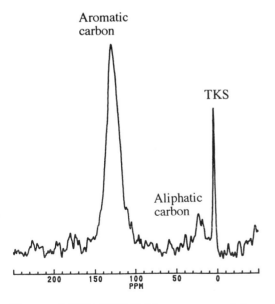

Figure 4 SPE ^{13}C NMR spectrum of the coke concentrate obtained after stripping the catalyst for 120 min.

The distribution of intensity in the aliphatic envelope in the ^{13}C NMR spectra for the coke from the second catalyst (Figure 2) is centred at ca 30 ppm and this indicates that CH_2 dominates over CH_3. In contrast, the aliphatic envelope in the spectrum of the first coke concentrate (Figure 1) contains a much larger contribution from CH_3 in the chemical shift range, 5-25 ppm (Figure 2). This difference may arise from the higher coke selectivity obtained with the second catalyst. A secondary factor could be that the lower flow rate used with the first catalyst has resulted in more side chain cracking as a consequence of the longer residence times of the reactive volatiles (particularly alkenes) in the reactor.

Stripped Cokes
The cokes obtained after stripping are highly aromatic with carbon aromaticities in excess of 0.90 (Figure 4 and Table 2). The EI mass spectrometry results (Figures 5 and 6) provide confirmatory evidence that these cokes are much more highly aromatic in character than the initial samples with the fragments from >2 ring polynuclear species being highly prominent in relation to the alkyl fragments (m/z 41, 43, 57, 71 and 85). Fragments from many 4-7 ring polynulcear species are evident (e.g 328 from 6 ring cata-condensed species). However, it is interesting to note that no material evolved under high vacuum in the mass spectrometer from the 120 minute sample until a much higher temperature had been reached (400 vs. 200°C).

The increase in aromaticity compared to the initial sample of ca 10 mole % carbon (Table 2) compares with the overall carbon loss of ca 15% obtained by stripping (Table 1, ca 1.5 to 1.3%). This suggests that side-chain cracking is probably the primary mechanism responsible for the formation of volatiles during stripping. However, the information from the DD ^{13}C NMR experiments on the initial and 120 min. stripped coke reveal that there has also been a growth in aromatic ring size (Table 2).

The non-protonated aromatic carbon concentrations obtained from the DD experiments have been corrected for alkyl substituted carbons (estimated from the aliphatic carbon contents) to leave the proportion of aromatic carbon which is in bridgehead positions (Table 2). Although the proportions of non-protonated aromatic carbon are similar for the initial and stripped cokes, the latter contains more highly condensed aromatic structures as suggested by the mass

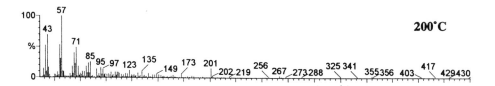

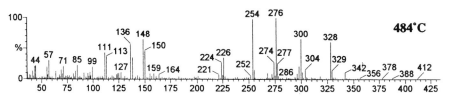

Figure 5 EI mass spectra of the coke concentrate obtained after stripping the catalyst for 60 min, the scans shown being taken at 200 and 484°C.

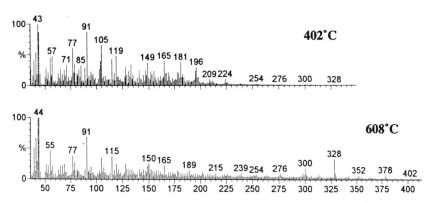

Figure 6 EI mass spectra of the coke concentrate obtained after stripping the catalyst for 120 min, the scans shown being taken at 402 and 608°C.

spectrometry results. Bridgehead carbon accounts for 45% of the aromatic carbon in the stripped coke which, on average, corresponds to approximately 8-10 ring peri-condensed structures. However, such structures are considerably less condensed than those found in the cokes from the vacuum residue and hydrogenated vacuum gas oil analysed previously (Snape *et al*, 1995, McGhee *et al*, 1995) where carbon in bridgehead positions accounted for over 50% of the aromatic carbon.

Table 2 Structural information from ^{13}C NMR analysis of the coked catalysts obtained from n-hexadecane

Catalyst	Aromaticity	C_{non-p}/C_{ar}	$C_{br.ar}/C_{ar}$
Initial, 6 dm^3 min^{-1}	0.82*	N.D.	N.D.
Initial, 4 dm^3 min^{-1}	0.83	0.55	0.40
Stripped for 120 min.	0.92	0.50	0.45

C_{non-p}/C_{ar} = fraction of aromatic carbon which is non-protonated (from DD experiments, $\pm$ 0.03).
$C_{br.ar}/C_{ar}$ = fraction of aromatic carbon in bridgehead positions ($\pm$ 0.04).
$C_{br.ar}/C_{ar} = C_{non-p}/C_{ar} - C_\alpha/C_{ar}$ where aliphatic carbon adjacent to an aromatic ring (C_α) has been assumed to be ca 70% of the aliphatic carbon based on proportions of aliphatic CH_2 and CH_3 (Love $et\,al$, 1993; Snape $et\,al$, 1995).
* = not including carboxyl and carbonyl .
N.D. = not determined.

As well as the increase in the degree of condenation of the aromatic groups, the aliphatic region in the ^{13}C NMR spectrum of the stripped coke indicates that CH_3 is the dominant aliphatic group (10-25 ppm) with no discernible peak at ca 30 ppm from CH_2 (Figure 4).

CONCLUSIONS

The demineralisation-quantitative ^{13}C NMR methodology successively used previously on FCC refinery catalysts has been extended to cokes at similar concentrations (ca 1% w/w) prepared from n-hexadecane in a laboratory-scale fluidised-bed reactor. The initial cokes possess considerable aliphatic character, but stripping for periods of up to 2 hours gives rise to an increase in aromaticity with the aromatic structure becoming more highly condensed. However, the average aromatic ring size (8-10 rings) for the stripped n-hexadecane coke is significantly lower than those found previously for heavy refinery feed cokes (10-20 rings). These latest findings provide further evidence that the structure of FCC coke is governed by the feed, even after prolonged stripping.

ACKNOWLEDGEMENTS

The authors thank the Engineering & Physical Sciences Research Council (EPSRC) for financial support of this work through (i) grant No. GR/H/24990, including a studentship for B.J. McGhee and (ii) the Mass Spectrometry Service Centre at University College of Swansea. Thanks also go to N.O.C and Ageco Oil Company, Libya for a scholarship for A. Aziz H. Mohammed.

REFERENCES

Butt, J.B. (1972). Catalyst Deactivation, Adv. Chem. Ser., 109, 259.

Dai, P-S.E., Neff, L.D. and Edwards, J.C. (1994). Effect of secondary porosity on gas-oil cracking activity, in Fluid Catalytic Cracking III, Am. Chem. Soc. Symp. Ser. No. 571 (Occelli, M.L. and O'Connor, P..eds.), p 69.

Durand B. and Nicaise G. (1980). Procedures for kerogen isolation, in Kerogen-insoluble organic matter from sedimentary rocks (Durand B. ed.), Editons Technip, Paris, p 35.

Groten W.A., Wojciechowski B.W. and Hunter B.K. (1990). Coke and deactivation II-formation of coke and minor products in the catalytic cracking of n-hexene on USHY zeolite, J. Catal., 125, 311.

Hughes, R.; Hutchings; G., Koon, C.L.; McGhee, B.; Snape, C.E. (1994). A fundamental study of the deactivation of FCC catalysts, in Catalyst Deactivation (Delmon, B. and Froment, G.F., eds.), Elsevier, 1994[a] and Studies in Surface Science and Catalysis, 88, 377.

Hughes R., Hutchings G., Koon C.L., McGhee B.J. and Snape C.E. (1994[b]). A comparison of the propensity of quinoline and phenanthrene to deactivate FCC catalysts, PREPRINTS, Div. of Petrol. Chem., ACS, 39 (3), 379.

Hughes, R., Hutchings, G., Koon, C.L., McGhee, B., Snape, C.E. and Yu, D. (1995). The effect of feedstock additives on FCC catalyst deactivation, PREPRINTS, Div. of Pet. Chem., Am. Chem. Soc., 40(3), 413.

Love G.D., Law R.V. and Snape C.E. (1993). Determination of the non-protonated aromatic carbon concentrations in coals by single pulse excitation ^{13}C NMR, Energy & Fuels, 7, 639.

McGhee, B., Snape, C.E., Andresen, J.M., Hughes, R., Koon, C.L. and Hutchings, G. (1995). Characterisation of coke from FCC catalysts by solid state ^{13}C NMR and mass spectrometry, PREPRINTS, Div. of Pet. Chem., Am. Chem. Soc., 40(3), 379.

O'Connor, P. and Pouwels, A.C., (1994). FCC catalyst deactivation: a review and directions for further research, in Catalyst Deactivation 1994 (Delmon, B. and Froment, G.F., eds.), Elsevier and Studies in Surface Science and Catalysis, 88, 129 and references therein.

Saxby, J.B. (1976) in Oil Shale (Yen T.F. and Chilingarian G.V., eds.), Elsevier, p 103.

Turlier, P., Forissier, M., Rivault, P., Pitault, I. and Bernard, J.R. (1994). Catalyst fouling by coke from vacuum gas oil in fluid catalytic cracking reactors, in Fluid Catalytic Cracking III, Am. Chem. Soc. Symp. Ser. No. 571, (Occelli, M.L. and O'Connor, P. eds.), p 98.

Snape, C.E., McGhee, B., Andresen, J., Hughes, R., Koon, C.L. and Hutchings, G. (1995). Characterisation of coke from FCC refinery catalysts by quantitative solid state ^{13}C NMR, Appl. Catal. A: General, 129, 125.

Wolf, E.H. and Alfani, A. (1982). Catalyst deactivation by coking, Cat. Rev. Sci. Eng., 24, 329 and references therein.

NEW FCC DEMETALLIZATION PROCEDURES

Stephen K. Pavel

Coastal Catalyst Technology, Inc., Houston, Texas

Fluid Catalytic Cracking is a valuable refining process to upgrade heavy hydrocarbons to high valued products (Avidan, 1993). Over 1,100 tons per day of Fluid Catalytic Cracking Unit (FCCU) catalyst is used worldwide in over 200 FCCUs. During the cracking reaction catalyst is contaminated by elements deposited from feedstocks which include nickel, vanadium, iron, et al. Contaminant elements are partitioned through the fluid catalytic cracking process. Carbon, sulfur, nitrogen, hydrogen, et al. are burned off in the regenerator. Some elements from feedstocks have been found on surfaces of the regenerator, and a portion of other remaining elements are carried with the products to fractionation. The bulk of contaminant elements from feedstock remain in the circulating catalyst system.

291

Through the cycles of cracking, fresh catalyst deactivation is caused by contaminant blockage of active sites by nickel, vanadium, iron, et al., and by steam catalyzed by contaminants, including vanadium, sodium, et al.(Pine, 1990; Pavel, 1992; Occelli, Gould, and Drake, 1994). To compensate for decreased FCC feedstock conversion and product selectivity, a portion of the circulating catalyst equilibrium inventory is withdrawn for spent catalyst disposal, and fresh catalyst is added to the system (Habib and Venuto, 1979).

Catalyst Demetallization (DEMET)

Demetallization processing (DEMET) takes a portion of spent FCC catalyst and removes a portion of metal contaminants by pyrometallurgical (calcining, sulfiding, nitrogen stripping, chlorinating) and hydrometallurgical (leaching, washing, drying) procedures, to return the base demetallized spent catalyst to the FCC. Standard demetallization (Elvin and Pavel, 1991) has most frequently utilized 787°C calcining and sulfiding with 343°C chlorinating. Off-gases from reactors are scrubbed. Contaminant metals are precipitated and filtered for disposal in the same manner used for spent catalyst, or they can be shipped to a Best Demonstrated Available Technology recycler of metals, depending on the client preference or regulatory environment. The operation of one unit for one refiner has resulted in the recycling of over 13,630 metric tons of spent FCC catalyst back to the FCC (Pavel and Elvin, 1996).

Demetallized spent FCC catalyst recycle has reduced the requirements for fresh catalyst additions and reduced generation of catalyst fines. DEMET processing removes contaminants known to be detrimental to conversion, product selectivities, and mechanical performance of the FCC. With DEMET capacity sized to reduce metals levels on circulating catalyst, yields could be improved due to lower metals on circulating catalyst.

New DEMET procedures

Standard DEMET processing utilizes a series of pyrometallurgical and hydrometallurgical procedures for metal removal. By removal of contaminants, access channels to active sites are renewed. Care is taken during processing to maintain the catalyst integrity and hydrothermal stability. New advancements in sulfidization and aqueous processing have further improved metal removal and demetallized spent catalyst characteristics of high initial activity, and low hydrothermal deactivation rates.

Catalyst deactivation

Fresh catalyst deactivation is covered directly and indirectly through most every paper and presentation on catalysis. Fresh catalyst is deactivated thermally, and hydrothermally (Occelli, Gould, and Drake, 1994). Catalysts are studied, modeled, and approached by various methods in laboratory testing (Peters, 1993). It is difficult to replicate commercial operations within the confines of laboratory environment, but many methods provide a relative direction for commercial decisions of catalyst selection and replacement rates of addition and withdrawal. Addition rates set the average age of the circulating catalyst.

For those units with very low metals in feedstock, the addition rates might be just sufficient to make up for losses through the reactor cyclones and regenerator cyclones. Activity of the catalyst is maintained in that equilibrium. On the other side of the spectrum, metals in feedstock might be so high that it is necessary to replace a high percentage of the circulating inventory to maintain an acceptable yield structure. The deactivation rates are higher for catalysts in units with higher temperatures, higher steam partial pressures, higher average age of circulating inventories, and higher contaminant metals. To understand the contaminant metals it is necessary to review the metal balance of the system including both feedstocks and fresh catalysts.

Feedstock contaminants

Feedstock contaminants are generally addressed individually by laboratory methods. A single contaminant element is deposited on catalysts by a variety of methods including impregnation and cyclic deposition. In certain cases, two elements might be co-deposited which provides some observation on the interaction of those elements. On the other hand, commercial FCCU feedstock contains a full spectrum of elements, not just nickel and vanadium. A recent summary of the impact of various contaminants included a more extensive list than most studies: V, Ni, Fe, Cu, Na, K, Mg, Ca, Ba, C, N (O'Connor et al., 1995).

Table 1 shows contaminant elements found in several different feedstocks. As the list of elements is not exhaustive, there are most likely other elements in FCCU feedstock and elements found in crude oil often concentrated in the heavy fractions (Filby, 1975). Testing for elements has been limited in commercial units due to the cost and time involved in accurate analyses. It is clear that nickel and vanadium are not the only contaminant elements. Particular contaminant levels in feedstock might be so low that measurement is difficult, but when contaminants are concentrated on a catalyst to the low levels found on spent catalyst, e.g. copper usually less than 50 ppm, they can contribute to site blockage and non-specific reactions degrading conversion and/or selectivity.

Lower cost higher metals feedstocks for the FCCU have been utilized to maximize refinery margins. Atmospheric resids, virgin or hydrotreated, are

Table 1. FCCU Feedstock Elemental Analyses (ppm) by ICP

Element	Feed1	Feed2	Feed3	Feed4	Feed5	Feed6	Feed7
Aluminum	2.9	38.7	0.4	1.6	4.8	nt	0.5
Antimony	0.2	2.1	1.1	0.2	0.2	<0.2	<0.2
Barium	nt	nr	0.9	<0.1	<0.1	nt	<0.2
Beryllium	nt	nt	nt	nt	nt	nt	nt
Bismuth	0.6	bdl	<0.1	0.2	0.2	<0.2	<0.2
Boron	<0.1	4.3	<0.1	0.7	<0.1	nt	<0.2
Calcium	19.9	15.3	1.6	2.7	0.5	0.4	0.5
Carbon	nr	nr	1.4	7.8	4.4	2.2	4.8
Cerium	0.2	nr	<0.1	<0.1	<0.1	nt	nt
Chromium	0.6	nr	<0.1	<0.1	<0.1	<0.2	<0.2
Copper	0.2	0.4	0.2	0.1	<0.1	<0.2	<0.2
Iron	14.3	16.6	5.2	16.0	7.6	3.4	5.3
Lanthanum	0.1	nr	<0.1	<0.1	<0.1	nt	nt
Lead	0.3	2.4	0.1	0.3	0.6	<0.2	<0.2
Magnesium	2.5	3.5	2.6	1.0	0.1	0.2	0.3
Manganese	0.2	0.1	<0.1	0.1	<0.1	<0.2	<0.2
Neodymium	<0.1	nr	<0.1	<0.1	<0.1	nt	nt
Nickel	2.2	4.8	3.4	22.0	3.9	13.0	12.1
Phosphorus	0.7	2.3	0.6	0.1	0.2	<0.2	<0.2
Potassium	0.6	5.7	0.6	0.5	0.7	nt	<0.2
Praseodymium	<0.1	nr	<0.1	<0.1	<0.1	nt	nt
Selenium	nt	bdl	<0.1	0.2	0.2	nt	<0.2
Silica	nt	nt	nt	nt	nt	nt	nt
Sodium	21.4	27.4	1.4	28.0	1.2	2.1	0.5
Strontium	0.6	nr	<0.1	<0.1	<0.1	<0.2	<0.2
Sulfur	nr	nr	1.3	1.2	0.3	0.13	0.37
Tin	0.1	0.4	0.7	0.5	0.7	<0.2	0.8
Titanium	0.1	0.7	<0.1	0.1	<0.1	<0.2	<0.2
Vanadium	3.5	4.3	2.4	40.0	6.5	0.9	8.6
Zinc	0.9	2.6	0.4	1.2	0.3	nt	<0.2
Zirconium	<0.1	nr	0.1	<0.1	<0.1	nt	<0.2

note: "nt" indicates "not tested" at the time of that sample
         "nr" indicates "not reported"

often incorporated in the FCCU feedstock blend.   Old units have been

modified, and new units have been constructed to take advantage of processing

higher percentages of heavier feedstocks.   As metals in feedstocks increase,

refiners make the trade-offs which balance lower product value yields at higher

metals versus additional fresh catalyst costs for lower metals and higher product

value yields.   Certain refiners have elected to maintain their circulating FCCU

catalyst at a higher metals level rather than increasing fresh catalyst additions

commensurate with the increased metals in the feedstock.

Contaminant effects on fresh FCC

Contaminant elements are deposited on the catalyst during the reaction step.

The performance of the FCCU catalyst system is diminished due to "deposition

of 1) all contaminants as they restrict or block access to acidic reaction sites

either on the surface micropores, mesopores, or macropores [48-50]; 2)

contaminants (K, Li, Na, V, et al.) which catalyze the steam destruction of

active sites [51-55]; 3) contaminants (Ca, Cu, Cr, Fe, Na, Ni, V, Zn et al.) which

cause non-selective yields of lower valued secondary reaction products, e.g.,

coke, hydrogen, methane, etc. [56-60]; 4) contaminants (Ca, Co, Cr, Fe, K, Li, Mg,

Mn, Mo, Na, V, W, Zn, et al.) which reduce the structural strength of the

catalyst components, either zeolite or matrix, or both [61-63];   and, 5) contaminants that deposit on surfaces within the regenerator (Al, Ba, C, Ca, Cr, Fe, K, Mg, P, S, Si, Zn et al.) [64] that cause mechanical plugging and unit shut-downs." (Pavel, 1995).

Alternatives to mitigate higher metals effects

Fresh catalyst and additive manufacturing companies responded to higher metals levels by introducing passivators and traps. Although some nickel passivators have been shown effective in reduction of dry gas made, certain passivators add additional elements to the circulating system which might complicate disposal in some locations. After promising results in the laboratory, a few vanadium traps have been introduced, but the commercial environment is somewhat different than the laboratory.

Traps and passivators are targeted for a particular contaminant (Ni or V). However, FCCU feedstock contains a wide range of contaminant elements (Ni, V, Fe, Ti, Na, et al.). In addition to traps diluting inventory of active sites available for catalytic cracking, controlling one element allows the other contaminant elements to increase. As the full range of contaminant elements increase, microporous channels with active sites are blocked. Rather than blocking active sites with additives, or diluting the active sites of the circulating catalyst inventory, demetallization removes the wide range of elements deposited from feedstocks, separating the contaminant (Ni, V, Fe, Ti, Na, et al.) from the catalyst base, renewing access to active sites.

Previous DEMET publications focused on full-scale applications. FCCUs at other locations have feedstocks with higher vanadium to nickel ratios. Many refiners have identified vanadium as a particular problem for their FCCU catalyst. To show a solution to the problem of higher vanadium levels, pilot plant demetallized spent FCCU catalyst, and MAT results after severe steaming show the effectiveness of metals removal and new technology development to improve the activity and stability of the demetallized catalyst compared to the original spent catalyst.

Elemental analysis of FCC

Elemental analyses of spent FCCU catalyst solids and leachates are necessary to characterize the total FCCU catalyst system including yield impacts and environmental considerations. Analysis of the solid spent and fresh catalyst are necessary to determine the elements available for removal from fresh and spent FCCU catalyst by demetallization. Although some consider just nickel and vanadium the influential contaminants, many of the other elements are known to be detrimental to catalysis.

Elements should not be ignored or assumed to have little effect, as very low levels of certain elements can be extremely influential. For example, platinum is maintained by additives, or incorporation into fresh catalyst compositions, to levels less than 1 ppm in the circulating inventory, and it effects the conversion of carbon monoxide to carbon dioxide, increases regenerator temperature, etc.. Many elements are used for other areas of catalysis (Trimm, 1980), and their influences might not have been yet investigated in relation to Fluid Catalytic Cracking.

Very seldom do investigations of zeolites, aluminas, catalysts, additives, etc. include a full description of the elements in the system, which shows the potential for many other catalytic effects from elements which were not detected.

Catalyst Leachate Analyses

Analysis of leachates provides an indication of the elements which will be removed during the demetallization leaching/washing procedures. Fresh, spent, and demetallized spent FCC catalysts were leached by various methods and in accordance with the Toxicity Characteristic Leaching Procedure (TCLP) EPA standard methods, and results were compared to proposed Land Disposal

Restrictions -- Phase II (LDRII) Universal Treatment Standards (UTS) (Federal Register, September 14, 1994). Demetallized catalyst TCLP leachates are below the UTS for all elements including vanadium. Additional leach tests were performed with deionized water flushed (DIF) through catalyst using 20 parts water to 1 part catalyst at two temperatures. DIF020 indicates a deionized water flush at 20°C; DIF100 indicates a deionized water flush at 100°C. The tests show that there are many more elements in the fresh and spent FCCU catalysts and their leachates.

Variation in chemical composition using different analytical techniqes are summarized in tables 2-8. Inductively Coupled Argon Plasma Spectroscopy (ICP) provides the fresh catalyst solid and leachate analysis (Table 2). Additional elements are deposited on the fresh catalyst from the feedstock resulting in spent catalyst. ICP provides the analysis of spent catalyst and it's leachate (Table 3). Following withdrawal from the regenerator, the spent catalyst is demetallized. ICP provides the demetallized spent catalyst solid and leachate analyses (Table 4). Xray Fluorescence Spectroscopy (XRF) provides additional data on the wide range of elements found in the FCC system: fresh (Table 5), after elemental deposition as spent (Table 6), and after demetallization (Table 7). Additional elements not analyzed but most likely present on either fresh, spent, or demetallized spent catalysts are also recognized (Table 8).

Table 2. Solid and Leachate Analyses (ppm) by ICP

Element	UTS LDR-II TCLP	Fresh Catalyst Solid	TCLP	DIF020	DIF100
Aluminum	---	142110	87.469	0.262	0.549
Antimony	2.1	246	bdl	bdl	bdl
Arsenic	5.0	215	0.223	bdl	bdl
Barium	7.6	118	0.018	bdl	bdl
Beryllium	0.014	3	0.006	bdl	0.002
Bismuth	---	bdl	bdl	bdl	bdl
Boron	---	20	0.029	0.009	0.021
Cadmium	0.19	4	bdl	bdl	bdl
Calcium	---	1372	7.059	0.279	0.278
Cerium	---	4682	13.183	bdl	0.085
Chromium	0.86	675	0.043	0.006	0.011
Cobalt	---	8	bdl	bdl	bdl
Copper	---	9	0.013	0.003	0.009
Iron	---	3053	0.219	0.011	0.016
Lanthanum	---	3518	8.401	0.009	0.017
Lead	0.37	27	bdl	bdl	bdl
Lithium	---	42	0.054	0.003	0.007
Magnesium	---	316	2.041	0.094	0.106
Manganese	---	12	0.055	bdl	0.001
Mercury	0.020	bdl	bdl	bdl	bdl
Molybdenum	---	8	0.016	bdl	0.011
Neodymium	---	2303	8.914	bdl	0.036
Nickel	5.0	33	0.049	bdl	0.006
Potassium	---	890	2.801	0.043	0.348
Praseodymium	---	587	1.991	bdl	bdl
Selenium	0.16	11	0.046	bdl	bdl
Silicon	---	310010	68.917	2.831	8.428
Silver	0.3	bdl	bdl	bdl	bdl
Sodium	---	3178	62.518	7.661	12.795
Sulfur	---	2943	84.625	30.348	61.931
Thallium	0.078	92	0.047	0.014	0.032
Tin	---	bdl	bdl	bdl	bdl
Titanium	---	4170	0.064	0.036	0.029
Vanadium	0.23	63	0.137	0.139	0.160
Zinc	5.3	91	bdl	bdl	bdl

Element	UTS LDR-II TCLP	Spent Catalyst			
		Solid	Leachates		
			TCLP	DIF020	DIF100
Aluminum	---	141190	114.813	1.574	2.961
Antimony	2.1	1526	1.766	3.207	9.491
Arsenic	5.0	250	1.495	bdl	bdl
Barium	7.6	52	0.667	0.008	0.014
Beryllium	0.014	38	0.368	0.309	0.509
Bismuth	---	7	0.319	0.011	0.015
Boron	---	20	0.109	0.116	0.094
Cadmium	0.19	20	bdl	bdl	bdl
Calcium	---	1819	1.889	0.269	0.283
Cerium	---	1171	14.915	bdl	0.065
Chromium	0.86	585	bdl	bdl	bdl
Cobalt	---	193	0.183	0.007	0.013
Copper	---	29	0.103	0.003	0.002
Iron	---	4292	0.328	0.152	0.301
Lanthanum	---	2757	46.975	0.162	0.246
Lead	0.37	bdl	bdl	bdl	bdl
Lithium	---	56	0.431	0.155	0.314
Magnesium	---	62	0.827	0.079	0.094
Manganese	---	12	0.052	bdl	0.001
Mercury	0.020	bdl	bdl	bdl	bdl
Molybdenum	---	14	0.227	0.263	0.474
Neodymium	---	1703	20.987	0.046	0.192
Nickel	5.0	3431	1.868	0.148	0.269
Potassium	---	242	1.205	0.242	0.224
Presodymium	---	432	6.347	bdl	bdl
Selenium	0.16	11	0.040	.bdl	bdl
Silicon	---	310485	50.384	3.828	32.506
Silver	0.3	bdl	bdl	bdl	bdl
Sodium	---	2283	26.993	8.416	16.473
Sulfur	---	446	1.134	0.591	0.882
Thallium	0.078	60	0.867	1.838	0.353
Tin	---	8	bdl	bdl	bdl
Titanium	---	7550	0.016	0.064	0.165
Vanadium	0.23	4967	48.437	41.162	68.498
Zinc	5.3	135	0.162	bdl	bdl

Table 3. Solid and Leachate Analyses (ppm) by ICP

Table 4. Solid and Leachate Analyses (ppm) by ICP

Element	UTS LDR-II TCLP	Demetallized Spent Catalyst — Solid	TCLP	DIF020	DIF100
Aluminum	---	175200	34.010	6.589	4.720
Antimony	2.1	310	bdl	bdl	bdl
Arsenic	5.0	96	1.480	0.113	0.205
Barium	7.6	77	0.329	0.042	0.060
Beryllium	0.014	10	0.012	0.003	0.004
Bismuth	---	bdl	bdl	bdl	bdl
Boron	---	14	0.067	0.043	0.037
Cadmium	0.19	30	bdl	bdl	bdl
Calcium	---	961	4.061	1.087	1.084
Cerium	---	1264	26.627	1.984	3.481
Chromium	0.86	579	bdl	bdl	bdl
Cobalt	---	12	0.142	0.040	0.052
Copper	---	9	0.120	0.002	0.001
Iron	---	902	2.320	0.069	0.053
Lanthanum	---	2782	57.195	5.492	8.928
Lead	0.37	bdl	bdl	bdl	bdl
Lithium	---	32	0.385	0.223	0.285
Magnesium	---	33	0.844	0.323	0.333
Manganese	---	8	0.036	0.008	0.012
Mercury	0.020	bdl	bdl	bdl	bdl
Molybdenum	---	4	0.058	bdl	bdl
Neodymium	---	1182	31.298	1.783	3.293
Nickel	5.0	153	1.745	0.378	0.547
Potassium	---	281	0.097	0.010	0.012
Presodymium	---	381	9.512	0.535	1.071
Selenium	0.16	bdl	bdl	bdl	bdl
Silicon	---	282500	55.571	2.001	8.440
Silver	0.3	bdl	bdl	bdl	bdl
Sodium	---	2211	16.367	10.051	12.147
Sulfur	---	2357	86.837	7.669	13.326
Thallium	0.078	bdl	bdl	bdl	bdl
Tin	---	bdl	bdl	bdl	bdl
Titanium	---	3800	0.264	0.039	0.016
Vanadium	0.23	1176	0.123	0.226	0.187
Zinc	5.3	95	1.360	0.036	0.098

Table 5. Other elements by XRF: Fresh Catalyst

Bromine		<	0.4
Cesium		<	7.5
Gallium	28.6	+/-	1.0
Germanium		<	0.7
Indium		<	0.6
Iodine	9.4	+/-	2.8
Niobium	13.7	+/-	1.0
Phosphorus	653.0	+/-	38.0
Rubidium	3.6	+/-	0.5
Silver		<	0.8
Strontium	60.5	+/-	1.0
Tantalum		<	4.9
Tellurium		<	1.4
Thorium	6.7	+/-	1.0
Tungsten		<	3.1
Uranium	2.7	+/-	1.0
Yttrium	59.8	+/-	1.0
Zirconium	103.0	+/-	1.0

Table 6. Other Elements by XRF: Spent Catalyst

Bromine		<	0.4
Cesium	22.3	+/-	10.0
Gallium	53.2	+/-	2.0
Germanium	2.1	+/-	0.5
Indium		<	1.8
Iodine	22.5	+/-	4.0
Niobium	25.3	+/-	1.0
Phosphorus	666.0	+/-	37.0
Rubidium	2.9	+/-	0.6
Silver		<	1.4
Strontium	67.5	+/-	1.0
Tantalum		<	8.0
Tellurium		<	2.5
Thorium	18.7	+/-	1.0
Tungsten		<	18.9
Uranium	4.1	+/-	1.1
Yttrium	32.6	+/-	1.0
Zirconium	125.0	+/-	2.0

Table 7. Other elements by XRF: Demetallized Spent Catalyst

Bromine			<	0.4	
Cesium	22.3		+/-	10.0	
Gallium	29.7		+/-	1.0	
Germanium			<	1.0	
Indium			<	1.8	
Iodine	10.3		+/-	3.0	
Niobium	9.9	+/-	0.8		
Phosphorus	604.0		+/-	31.0	
Rubidium	1.7		+/-	0.6	
Silver			<	1.4	
Strontium	60.6		+/-	1.0	
Tantalum			<	8.0	
Tellurium			<	2.5	
Thorium	16.8	+/-	1.0		
Tungsten			<	18.9	
Uranium	3.2	+/-	1.1		
Yttrium	27.4		+/-	1.0	
Zirconium	125.0		+/-	2.0	

Table 8. Elements not tested in study but often found in feedstocks and on catalysts

Other elements not in testing but most likely present:	Other Rare Earths Present in Samples	Other Elements In Crude/Catalysts
Carbon	Samarium	Fluorine
Chlorine	Europium	Iridium
Hafnium	Gadolinium	Osmium
Hydrogen	Terbium	Palladium
Nitrogen	Dysprosium	Platinum
Oxygen	Holmium	Rhenium
Scandium	Erbium	Rhodium
	Terbium	Ruthenium
	Ytterbium	
	Lutetium	

Notes on elemental analysis

Elemental analyses of solids were performed using ICP and XRF; elemental analyses of leachates were performed by ICP. All tests were performed at third party laboratories. When an element was below detection limit of the ICP it is shown as "bdl". When an element was not included in the third party testing it is shown as "nt" - not tested. By request, certain results are not recorded and are shown as "nr". Elements in spent catalyst are most often higher than the fresh catalyst. Some elements of spent appear lower than the fresh catalyst. This occurs when there are no additional deposits of that element from the feedstock and as a result the original elements from the fresh catalyst is diluted by deposited elements.

Analysis of the demetallized spent catalyst represents just one of the special demetallization procedures recently developed. The metals content shows metal removal, but it should be noted that as various elements are removed, those which are not removed will appear as an increase in the material. The analysis also reflects the use of a particular wash water system which has been supplanted in more recent procedures. The particular interest in the analysis of demetallized spent catalyst TCLP leachate is that the leachate elements are reduced in comparison to spent catalyst leachate and all elements are below the proposed Universal Treatment Standards.

Spent Catalyst Disposition Alternatives

Spent FCCU catalyst disposal quantities have been published in various formats, and disposition alternatives include those with DEMET options which recover metals:

1)   On-site demetallization and recycle for FCCU catalysis -- on-site source reduction;

2)   Off-site demetallization and recycle for FCCU catalysis -- original application recycle;

3)   Off-site spent catalyst (ultra-low metals) sale to others for further metals loading, a limited market at equilibrium which requires disposal by others and does not affect disposal of total replacement volumes of fresh catalyst sold, and the metals deposited from feedstocks;

4)   Off-site demetallization for metals recovery prior to secondary use or disposal;

5)   Other waste treatment technologies prior to disposal, e.g. solidification, stabilization, vitrification, cementation, etc.;

6)   Cement kilns, with or without pretreatment, either 6a) cement kilns permitted/licensed for hazardous wastes, or 6b) cement kilns blending wastes as alternative feed stocks;

7)   Landfills, either a) landfills permitted/licensed for hazardous wastes, or b) landfills not permitted for hazardous waste.

Demetallized spent FCC

Demetallization for primary recycling to FCCUs can reduce fresh catalyst additions, reduce circulating inventory metals (or hold a metals level with increasing feed stock metals), and reduce leachable metals for a portion of spent catalyst withdrawn.

Contaminant metals are controlled in FCCUs due to their deleterious affects on conversion, selectivity, and deactivation of fresh catalyst. Similarly, cement kiln blends are limited to a kiln blend of 83 ppm vanadium due to potential problems with refractories (Petrovsky, 1994). For those kilns not already at a vanadium limit due to vanadium content of the local quarry supply, the dilution of vanadium sets the limit of spent FCCU catalyst processed to 1-2% of kiln feed stock. The mobility of vanadium is well known in refining, and it appears the conditions of cement kiln processing do not limit vanadium mobility. Demetallization could be used to reduce the leachable metals prior to secondary recycling (cement kilns, etc.) or disposal. Using DEMET to remove vanadium from feedstocks to cement kilns would increase capability to substitute spent FCCU catalyst in the kiln blend. (Pavel, 1995)

Improved metals removal

Special DEMET procedures were developed by rigorous thermodynamic modeling of the sulfidization environment to ensure the reaction mix and temperature appropriate to convert available contaminant oxides to sulfides. Accurate analyses of all contaminant elements (not just nickel and vanadium) are required for accurate modeling. Improved metals removal enables more efficient utilization of hardware and provides the ability to lower metals on circulating catalyst. Improved metals removal indicates that more sites have been uncovered, and micropore channels leading to active catalytic sites restored.

Improved hydrothermal stability

Special DEMET procedures were developed by rigorous modeling of solution properties (Pourbaix, 1974) during the leach and wash step. Additional proprietary steps are incorporated in the leaching and washing steps prior to drying on the belt filter. Aqueous processing modifications resulted in a number of attractive alternatives which can be selected through regulating variables in a single flexible design DEMET unit. For simplicity of graphic and tabular presentation, only one of the alternative advanced procedures is shown, labeled "special." Spent FCC catalyst was demetallized in the laboratory pilot plant.

Catalyst performance testing options

There are multitudes of MAT and steaming conditions which have been developed for testing fresh and spent catalysts. Particular companies prefer testing fresh catalyst with the relatively more severe level of 100% steam, 787°C to provide indication as to which fresh catalyst is more hydrothermally stable. There are many tests other than steaming and MAT which might be utilized to test alternative catalysts: fixed fluid beds, FCCU pilot plant, modified MATs, cyclic deactivation and MAT, etc.. Previous testing at a corporate R&D facility using a wide range of equipment provided virtually the same results as indicated by their own steaming and MAT. If a catalyst can survive the severity of temperature, steam, and time, it would be expected to perform relatively the same in cyclic deactivation or pilot plant testing. Additional testing at other corporate R&D facilities have shown demetallized catalyst to be more coke selective than shown by MATs. For the purpose of comparing demetallized catalyst to spent and fresh catalysts, steaming and MAT tests were performed by a third party testing facility.

Microactivity Testing

Steaming, microactivity and XRD testing were performed in accordance with ASTM standards at third party laboratories. Steaming was performed at 787°C 100% steam for 4 hours and 16 hours. The results prior to steaming, then after 4 and 16 hours of steaming are provided in Tables 9, 11, 12 and 13 with a comparison of the fresh, spent, standard demetallized, and special demetallized sample performance as shown by MAT testing. Figures following provide graphic support for several variables. Table 9 shows the improvement in metal removal as the temperature of sulfidization is increased from 787 to 843°C.

Full yield MAT tests were performed on the fresh, equilibrium and demetallized catalysts prior to steaming and after each was steamed for 0, 4, and 16 hours with 100% steam at 787°C. Again, there are more tests that can be performed, e.g. steaming for 0, 0.5 1, 2, ... hours, to better define the inflection point of the initial effect of steaming prior to the MAT after four hours. As the critical factor was stability after 16 hours of steaming, the 0, 4, 16 hour steaming followed by MAT provided ample data to make an accurate determination of fresh, spent, and demetallized catalyst resistance to hydrothermal deactivation.

Summary and discussion of results

Special DEMET procedures are shown superior to standard DEMET by results of steaming and MAT testing as demetallized catalyst activity and hydrothermal stability improves with special processing. Discussion of test results follow while details are provided in tables and figures described.

Table 9 shows standard demetallization increases as sulfidization temperature increases from 787°C to 843°C. Nickel removal is ~95% for both; vanadium

Table 9. Fresh, Equilibrium, and Demetallized Equilibrium Catalysts,
comparison of demetallization at 787 and 843°C sulfidization

	Fresh without DEMET	ECAT without DEMET	Standard 787°C DEMET	Standard 843°C DEMET
Metal analysis by XRF summary, ppm				
Nickel	26	4610	269	193
Vanadium	60	5790	1880	882
Copper	5	22	9	6
Iron	3650	4990	2330	1150
Titanium	5010	7880	3350	3230
Metal removal from base ECAT, wt%				
Nickel			93.58%	95.81%
Vanadium			67.53%	84.77%
Copper			58.99%	72.81%
Iron			53.31%	76.95%
Titanium			57.49%	59.01%
MAT yields, Conversion, wt%				
Before steaming	94.28	55.55	67.74	64.40
After 4 hours steam	64.53	48.73	52.06	51.34
After 16 hours steam	62.91	35.86	40.38	46.65
Deactivation Rate, Conversion loss wt%/hr				
from 0 to 4 hours steam	7.44	1.71	3.27	3.92
from 4 to 16 hours steam	0.13	1.07	0.97	0.39
from 0 to 16 hours steam	1.96	1.23	1.71	1.11

Figure 1 shows improved MAT conversion resulting from demetallization

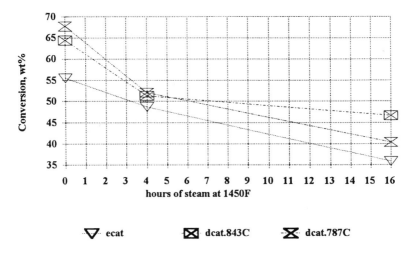

Figure 1. Conversion improved by Standard Demet at 787°C and 843°C Sulfidization.

removal increases from 67% to 85%; and iron removal increases from 53% to 77%. The MAT conversion after 16 hours of steaming is higher when catalysts were sulfided at 843°C rather than at 787°C, indicating that the deactivation rate is lower. Table 10 provides a metals removal summary: fresh, equilibrium, standard and special DEMET. Metals removal of standard and special procedures are comparable at this temperature: 95% nickel, 84% vanadium, 34% sodium, 76% iron. Tables 11, 12, 13 provide MAT details for no steam, after four and sixteen hours of steam. The effect of the special DEMET process on equilibrium MAT yields compared for fresh catalyst, ECAT, standard and special DEMET.

Table 10.   Fresh, Equilibrium, and Demetallized Equilibrium Catalysts,
comparison of standard and special demetallization procedures

	Fresh without DEMET	ECAT without DEMET	ECAT after Standard DEMET	ECAT after Special DEMET
Metal analysis by XRF summary, ppm				
Nickel	26	4610	269	303
Vanadium	60	5790	1880	959
Barium	139	196	150	152
Calcium	564	454	435	374
Chromium	29	99	52	51
Cobalt	49	233	67	58
Copper	5	22	9	9
Iron	3650	4990	2330	1120
Lead	27	38	23	10
Potassium	826	516	460	402
Titanium	5010	7880	3350	3340
Zinc	33	80	51	43
Metal removal from base ECAT, wt%				
Nickel			93.58%	93.43%
Vanadium			67.53%	83.44%
Barium			23.47%	22.45%
Calcium			4.19%	17.62%
Chromium			30.18%	48.29%
Cobalt			71.24%	74.94%
Copper			59.05%	59.05%
Iron			53.31%	77.56%
Lead			39.84%	73.70%
Potassium			10.85%	22.09%
Titanium			57.49%	57.61%
Zinc			36.02%	46.58%

Detailed MAT results are in the following tables 11 - 14, and Figures 2 - 10

Table 11. Fresh, Equilibrium, and Demetallized Equilibrium Catalysts,
catalyst quality before steaming, MAT, wt%

MAT yields before steaming	Fresh without DEMET	ECAT without DEMET	ECATafter Standard DEMET	ECATafter Special DEMET
Conversion	94.29	55.55	67.74	70.78
Kinetic Conversion	16.50	1.25	2.10	2.42
Gasoline/Conversion	0.27	0.63	0.56	0.54
Kin.Conv./Coke	0.79	0.15	0.24	0.21
Gasoline/Coke	1.22	4.11	4.36	3.32
Dry Gas/Kin. Conv.	0.38	1.55	1.23	1.03
LCO/LCO+Slurry	0.47	0.52	0.60	0.63
H2	0.29	0.46	0.26	0.46
C1	2.50	0.60	0.98	0.74
C2	1.94	0.56	0.62	0.62
C2=	1.78	0.77	0.97	1.13
DryGas (C1+C2s)	6.22	1.93	2.57	2.49
C3	17.06	0.85	2.30	1.74
C3=	2.17	2.74	4.34	4.84
Total C3s	19.23	3.59	6.64	6.58
IC4	14.36	2.28	6.17	5.60
NC4	6.65	0.56	1.35	1.22
Isobutene	0.39	0.93	0.96	1.17
Total butenes	1.20	3.24	3.91	4.51
Total C4s	22.21	6.08	11.43	11.33
Total LPG	41.44	9.67	18.07	17.91
GSL (C5-221°C)	25.41	34.98	38.10	38.34
LCO (221-343°C)	2.66	22.94	19.36	18.55
SLURRY (343°C+)	3.05	21.51	12.91	10.68
COKE	20.91	8.51	8.74	11.56

Table 12. Fresh, Equilibrium, and Demetallized Equilibrium Catalysts, catalyst quality after 4 hours steaming, MAT, wt%

MAT yields after 4 hrs 100% steam at 787°C	Fresh without DEMET	ECAT without DEMET	ECATafter Standard DEMET	ECATafter Special DEMET
Conversion	64.53	48.33	52.06	62.10
Kinetic Conversion	1.82	0.94	1.06	1.63
Gasoline/Conversion	0.63	0.61	0.68	0.63
Kin.Conv./Coke	0.37	0.07	0.14	0.45
Gasoline/Coke	6.66	3.68	9.33	7.24
Dry Gas/Kin. Conv.	1.44	1.89	1.72	1.24
LCO/LCO+Slurry	0.58	0.46	0.51	0.55
H2	0.08	0.72	0.30	0.21
C1	0.32	0.63	0.60	0.64
C2	0.63	0.55	0.63	0.60
C2=	0.77	0.60	0.59	0.79
Dry Gas (C1+C2s)	2.62	1.78	1.82	2.03
C3	1.63	0.63	0.76	1.13
C3=	4.19	2.49	3.05	4.40
Total C3s	5.82	3.12	3.81	4.53
IC4	4.64	1.29	2.35	2.54
NC4	1.04	0.36	0.47	0.51
Isobutene	1.00	1.19	1.32	1.62
Total butenes	4.28	3.41	3.98	5.45
Total C4s	9.96	5.06	6.80	8.50
Total LPG	15.78	8.18	10.61	12.03
GASOLINE	40.56	29.60	35.53	39.18
LCO	20.67	23.61	24.26	20.70
SLURRY	14.79	28.07	23.68	17.21
COKE	6.09	8.05	3.81	5.41

Table 13. Fresh, Equilibrium, and Demetallized Equilibrium Catalysts, catalyst quality after 16 hours steaming, MAT, wt%

MAT yields after 16 hrs 100% steam at 787°C	Fresh without DEMET	ECAT without DEMET	ECAT after Standard DEMET	ECAT after Special DEMET
Conversion	62.91	35.86	40.38	56.56
Kinetic Conversion	1.70	0.56	0.68	1.30
Gasoline/Conversion	0.66	0.56	0.65	0.68
Kin.Conv./Coke	0.37	0.07	0.14	0.45
Gasoline/Coke	9.09	2.44	5.43	13.45
Dry Gas/Kin. Conv.	1.11	2.98	2.55	1.37
LCO/LCO+Slurry	0.68	0.33	0.37	0.52
H2	0.07	0.64	0.34	0.21
C1	0.58	0.51	0.88	0.56
C2	0.60	0.59	0.47	0.57
C2=	0.70	0.57	0.37	0.66
Dry Gas (C1+C2s)	1.88	1.67	1.73	1.78
C3	1.19	1.01	0.66	0.79
C3=	4.21	1.73	2.21	3.97
Total C3s	5.40	2.74	2.87	4.76
IC4	3.94	0.54	0.95	2.54
NC4	0.83	0.19	0.28	0.51
Isobutene	1.17	0.63	1.19	1.99
Total butenes	4.77	1.63	3.17	5.70
Total C4s	9.54	2.36	4.40	8.75
Total LPG	14.94	5.11	7.87	13.51
GASOLINE	41.47	20.18	26.22	38.20
LCO	22.47	20.89	21.92	22.47
SLURRY	15.73	43.25	37.70	20.98
COKE	4.56	8.28	4.83	2.87

Table 11 shows the results from MAT prior to steaming.   The unsteamed

product yields from the fresh catalyst demonstrate poor selectivity, with high

coke, high gas and low gasoline.   When fresh catalyst is added to an FCCU in

large quantities in a resid operation, these yields are realized.   Depending on the

catalyst addition system, the size of the regenerator, the percentage of inventory

represented by the fresh catalyst addition, FCCU stack treatment (ESP, wet

scrubber, ...), etc., the effects of fresh catalyst additions are often very apparent

Table 12 shows results from MAT after four hours of 100% steaming at 787°C.

Table 13 shows results from MAT after sixteen hours of 100% steaming at

787°C. The special DEMET process catalyst after 16 hours of steaming yielded:

90% more gasoline than ECAT	and	16% more gasoline than standard DEMET
370% more IC4 than ECAT	and	128% more IC4 than standard DEMET;
250% more C4= than ECAT	and	48% more C4= than standard DEMET;
8% more LCO than ECAT	and	3% more LCO than standard DEMET;
51% less slurry than ECAT	and	34% less slurry than standard DEMET;
65% less coke than ECAT	and	5% less coke than standard DEMET.

Table 14 shows the effect of the special DEMET process on equilibrium

deactivation rates compared for fresh catalyst, ECAT, standard and special

DEMET.   The deactivation rate of the special demetallized catalyst is one half

Table 14. Fresh, Equilibrium, and Demetallized Equilibrium Catalysts,
Deactivation and Activity Recovery Calculations, MAT, wt%

100% steam at 787°C	Hours Steam	Fresh without DEMET	ECAT without DEMET	ECAT after Standard DEMET	ECAT after Special DEMET
Conversion	0	94.29	55.55	67.74	70.78
Conversion	4	64.53	48.33	52.06	62.10
Conversion	16	62.91	35.86	40.38	56.56
Deactivation rate (wt%/hour)					
from 0 to 4 hours steam		7.44	1.71	3.27	2.17
from 4 to 16 hours steam		0.13	1.07	0.39	0.46
from 0 to 16 hours steam		1.96	1.23	1.11	0.89
Conversion	0	0.00	38.73	29.88	26.54
Deactivation from	4	0.00	15.80	13.19	12.47
Fresh minus other	16	0.00	27.05	16.26	22.53
Activity above ECAT	0	38.73	0.00	8.85	12.19
MAT Conversion	4	15.80	0.00	2.61	3.33
Other minus ECAT	16	27.05	0.00	10.79	4.52
Activity recovery as	0	100%	0%	22.85%	39.32%
a percent of difference	4	100%	0%	16.52%	84.56%
Fresh minus ECAT	16	100%	0%	39.89%	76.52%
Average activity recovery		100%	0%	26.42%	66.80%

that of fresh catalyst, significantly less than the deactivation rate of ECAT, and less than standard demetallized catalyst. One might assume that the maximum possible degree of reactivation would increase the equilibrium catalyst MAT conversion to that of the fresh catalyst; the reactivation from demetallization is given as a percentage of this amount. It is recognized that this is a simplistic relative comparison, as most fresh catalysts are designed to deactivate to an equilibrium level.

Metal contamination blocks pores restricting access to active sites, and catalyze the steam deactivation of the fresh catalysts resulting in a spent equilibrium catalyst activity somewhat lower than the deactivated fresh catalyst, and deactivation rate higher than the fresh catalyst itself. The special demetallization process recovered 67% of the maximum possible activity compared to 26% for the standard DEMET. The combination of low deactivation rates and higher MAT conversions indicates that the special DEMET will be superior to the standard DEMET in commercial applications, resulting in lower fresh catalyst requirements to maintain FCCU activity.

Deactivation Rates and Activity Recovery

Deactivation rate can be calculated by a variety of methods. For the simple purposes of comparing fresh, spent and demetallized spent catalyst, the deactivation rate is calculated as the loss of conversion per hour of steaming.

Benefit of DEMET

Demetallization is thermodynamically favored at $787^{\circ}C$ and $843^{\circ}C$. MAT

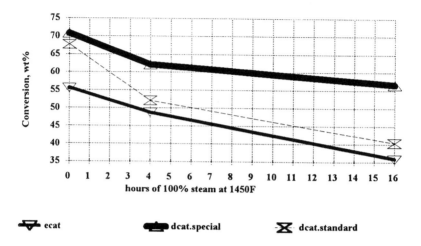

Figure 2. Special Procdures Increase Initial Activity and Hydrothermal Stability.

conversion from standard demetallization at both temperatures is shown in figure 1. Special demetallization procedures significantly improve the initial activity and hydrothermal stability as shown in figure 2. Figure 3-10 show the improvements gained by special demetallization procedures by increased C5+ Gasoline (figure 3), decreased Coke (figure 4), increased Gasoline/Coke Selectivity (figure 5), increased Kinetic Conversion (figure 6), increased Dynamic Activity = Kinetic Conversion/Coke (figure 7), decreased C2minus Gas/Kinetic Conversion (figure 8), increased Gasoline/Conversion Selectivity (figure 9), and increased LCO Selectivity = LCO/LCO+DCO (figure 10).

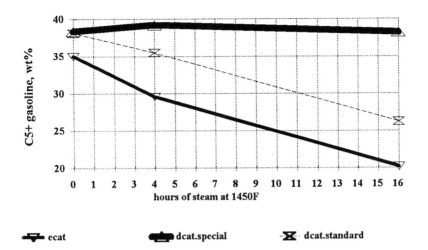

Figure 3. C5+ Gasoline Yield Increased by Std. Demet and Special Procedures.

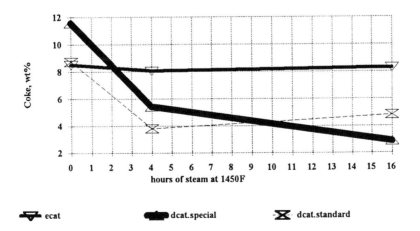

Figure 4. Coke Yield Reduced by Standard Demet and Special Procedures.

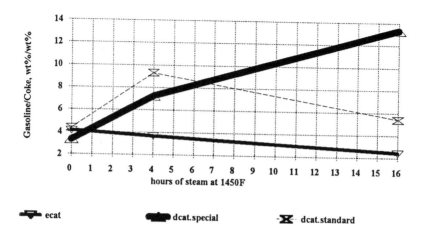

Figure 5. Gasoline / Coke Selectivity Increased by Special Procedures.

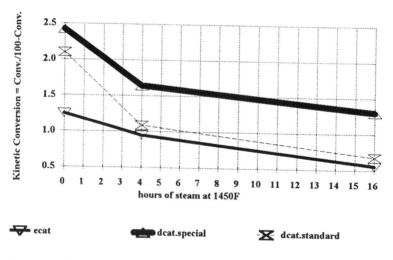

Figure 6. Kinetic Conversion Increased by Special Procedures.

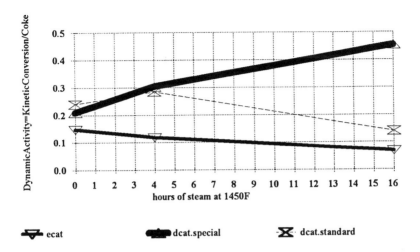

Figure 7. Kinetic Conversion / Coke Selectivity Increased by Special Procedures.

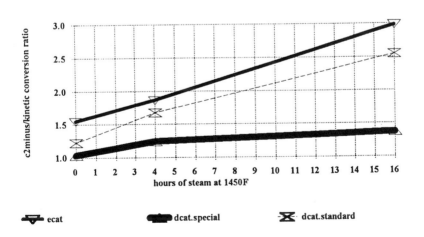

Figure 8. DryGas/Kinetic Conversion Reduced by Std. Demet & Special Procedures.

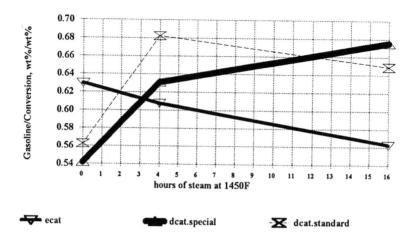

Figure 9. Gasoline Selectivity Increased by Std. Demet and Special Procedures.

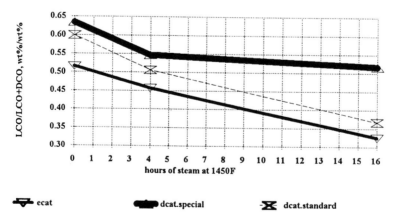

Figure 10. LCO Selectivity Increased by Special Procedures.

Conclusions

Demetallization reduces the leachability of contaminant elements measured by TCLP and de-ionized water flushes. After demetallization of spent FCCU catalyst all TCLP leachate elements are below the proposed Universal Treatment Standard levels.

For every indicator of catalyst performance after severe steaming, standard demetallization appears superior to equilibrium catalyst without demetallization, and special demetallization appears superior to standard demetallization and far superior to equilibrium catalyst without demetallization. The figures show higher conversion, and higher gasoline yields with lower coke for special demetallized catalyst compared to equilibrium catalyst. The figures also show the special demetallized catalyst performance ratios are also superior with lower dry gas/kinetic conversion ratios, higher gasoline/coke ratios, higher gasoline/conversion selectivities, higher light cycle oil selectivities, higher dynamic activities, as well as increased C3 and C4 olefins and isobutylene. DEMETallization appears very well suited for the marketplace of new gasolines and higher middle distillate demands.

The special DEMET process significantly increases the activity, selectivity and hydrothermal stability of the catalyst and minimizes the fresh catalyst required to maintain FCCU activity.

Acknowledgments

CCTI gratefully acknowledges our research associates at various locations, companies who have provided ample samples for process development, and our colleagues working to bring DEMET technology to refiners throughout the world. The authors appreciate and acknowledge the previous contributions to the fields referenced in this paper. Coastal Catalyst Technology, Inc., has already been recognized by the State of Texas for the effort in the reduction of solid waste (Texas Corporate Recycling Council, 1992).

References

Avidan, A. A. (1993). Origin, development and scope of FCC catalysis, J. S. Magee and M. M. Mitchell, Jr., Eds., Fluid Catalytic Cracking: Science and Technology, Studies in Surface Science and Catalysis, Vol. 76, (Amsterdam, Elsevier) 1.

Batchelder, M. L., ed., 1989 NPRA Q&A Session on Refining and

Petrochemical Technology, (Tulsa, OK: Gerald L. Farrar & Associates, Inc.), 58.

Batchelder, M. L., ed., 1993 NPRA Q&A Session on Refining and Petrochemical Technology, (Tulsa, OK: Gerald L. Farrar & Associates, Inc.), 53, 166.

Beguin, B., Garbowski, E., and Primet, M. (1991). Stabilization of Alumina toward Thermal Sintering by Silicon Addition, Journal of Catalysis, Vol. 127, 595.

Burtin, P., Bunelle, J. P., Pijolat, M., and M. Soustelle, M., 1987). Influence of Surface Area and Additives on the Thermal Stability of Transition Alumina Catalyst Supports. I: Kinetic Data, Applied Catalysis, 34, 225.

Cimbalo, R. N., Foster, R. L., and Wachtel, S. J. (1972). Deposited Metals Poison FCC Catalyst, Oil and Gas Journal (May 15), 112.

Clark, F. T., Springman, M. C., Willcox, D., and Wachs, I. E. (1993a). Interaction in Alumina-Based Iron-Vanadium Catalysts Under High Temperature Calcination and SO_2 Oxidation Conditions, Journal of Catalysis, 139, 1.

Clark, F. T., Springman, M. C., Willcox, D., and Wachs, I. E. (1993b). Interaction in Alumina-Based Iron-Vanadium Catalysts Under High Temperature Oxidation Conditions, <u>ACS Symposia Preprints</u>, vol. 38, No.3 (Washington, D. C.: ACS) 615.

Connor, Jr., J. E., Rothrock, J. J., Birkhimer, E. R., and Leum, L. N. (1957). Fluid Cracking Catalyst Contamination, Some Fundamental Aspects of Metal Contamination", <u>Industrial and Engineering Chemistry</u>, Vol. 49, No. 2, 276.

Elvin, F. J. (1990). Long Term effects of FCCU Catalyst Demetallization, presented at the Summer 1990 Meeting of AIChE, San Diego, California.

Elvin, F. J. and Pavel, S. K. (1991). Fluid Cracking Catalyst Demetallization - Commercial Results, <u>NPRA Annual Meeting</u>, #AM-91-40.

Elvin, F. J. and Pavel, S. K. (1993). Commercial Operations of a DEMET Unit, <u>NPRA Annual Meeting</u> #AM-93-54.

<u>Environment Reporter</u>, (1992). Appendix II--Method 1311 Toxicity Characteristic Leaching Procedure (TCLP). (WDC.: Bureau of National Affairs, Inc.), 310.

Federal Register (1994), <u>40 CFR Parts 148, et al. Land Disposal Restrictions Phase II -- Universal Treatment Standards, and Treatment Standards for Organic Toxicity Characteristic wastes and Newly Listed Wastes; Final Rule</u>, Sept. 19, 1994 , p. 47982.

Filby. R. H. (1975). The Nature of Metals in Petroleum. Yen, T. F., ed., . <u>The Role of Trace Metals in Petroleum</u>, (Ann Arbor, MI: Ann Arbor Science Publishers, Inc.), 31.

Fritz, P. O., and Lunsford, J. H. (1989). The Effect of Sodium Poisoning on Dealuminated Y-type Zeolites, <u>Journal of Catalysis</u>, 118, 85, 12.

Gossett, E. C. (1960). When Metals Poison Cracking Catalyst, <u>Petroleum Refiner</u>, (June) 177.

Habib, Jr., E. T., Owen, H., Synder, P. W., Streed, C. W., and Venuto, P. B. (1977). Artificially Metals-Poisoned Fluid Catalysts. Performance in Pilot Plant Cracking of Hydrotreated Resid, <u>Industrial & Engineering Chemistry, Product Research & Development</u>, Vol. 15, No. 4 , 291.

Habib, T. H. and Venuto, P. B. (1979). <u>Fluid Catalytic Cracking with Zeolite Catalysts</u>, (New York, NY: Marcel Dekker).

Kumar, R., Cheng, W.-C., Rajagopalan, K., Peters, A. W., and Basu, P. (1993). The effect of exchange cations on Acidity, Activity, and Selectivity of Faujasite Cracking Catalysts, Journal of Catalysis, 143, 595.

Mauge, F., Gallezot, P., Courcelle, J.-C., Englehard, P.,and Grosmangin, J., (1986). Hydrothermal aging of cracking catalysts. II. Effect of steam and sodium on the structure of La-Y zeolites, Zeolites, Vol. 6, 261.

Mitchell, B. R. (1980). Metals Contamination of Cracking Catalysts. 1. Synthetic Metals Deposition on Fresh Catalysts", Industrial Engineering Chemistry Product Research and Development, Vol. 19, 209.

Moorehead, E. L., McLean, J. B., and Cronkright, W. A. (1993). Microactivity evaluation of FCC catalysts in the laboratory: principles, approaches, and applications, J. S. Magee and M. M. Mitchell, Jr., Eds., Fluid Catalytic Cracking: Science and Technology, Studies in Surface Science and Catalysis, Vol. 76, (Amsterdam, Elsevier) 223.

Occelli, M. L., Gould, S. A. C., and Drake,B. (1994).   Atomic Force Microscopy Examination of the Topography of a Fluid Cracking Catalyst Surface. M. L. Occelli and P. O'Connor, Eds. Fluid Catalytic Cracking III Materials and Processes, ACS Symposium Series 571. (Washington, DC) 271.

Occelli, M. L., Gould, S. A. C., Baldiraghi, F. And Leoncini (1996). Atomic Force Imaging of the Surface of a Fluid Cracking Catalyst before and after Thermal and Hydrothermal Treatments. Advances in FCC Conversion Catalysts Symposia Preprint, Division of Petroleum Chemistry, ACS meeting, New Orleans, March 14-19, 1996. (Washington, DC., ACS), 377 and presentation.

O'Connor, P., Brevoor, E. Pouwels, A. C., and Wijnards, H. N. (1995). Catalyst Deactivation in Fluid Catalytic Cracking a Review of Mechanisms and Testing Methods. Catalysis in Petroleum Refining and Petrochemicals, Proceedings of the 5th Annual Workshop at the KFUPM Research Institute, Khahran, Saudi Arabia, p. 152.

O'Donoghue, E., and Barthomeuf, D. (1986). Specific influence of alkali-metal cations on OH groups and active sites in Y zeolite, Zeolites, Vol. 6, 267-271.

Pavel, S. K. (1992). The deactivating effects of vanadium and steam are compared for fresh, equilibrium, and demetallized catalyst. AIChE Annual Meeting, Miami, FL.

Pavel, S. K., and Elvin, F. J. (1993). Minimization of Petroleum Refinery Waste by Demetallization and Recycling of Spent FCCU Catalyst, Hager, J., Hansen,

B., Imrie, W., Pusatori, J., and Ramachandran, V., eds., Extraction and Processing for the Treatment and Minimization of Wastes, The Minerals, Metals & Materials Society, Washington, D.C., 1015.

Pavel, S. K. and Elvin, F. J. (1994). Elemental Analysis of FCCU catalysts: Fresh, Equilibrium, and Demetallized, Symposium on Recent Advances in FCC Technology, AIChE Spring National Meeting, Atlanta, GA.

Pavel, S. K. and Elvin, F. J. (1995). Fresh, Spent and Demetallized Spent FCCU Catalysts Characterized by Detailed Elemental Analysis of Powder and Leachate", Symposium on New Techniques in Materials and Catalyst Characterization Preprints, ACS, Division of Petroleum Chemistry, Washington, D.C., 198.

Pavel, S. K. and Elvin, F. J. (1996). New FCC DEMET Process to Produce Demetallized Spent Catalyst with High Activity and Low Hydrothermal Deactivation Rates. NPRA Annual Meeting #AM-96-45.

Peters, A. W. (1993). Instrumental methods of FCC catalyst characterization, J. S. Magee and M. M. Mitchell, Jr., Eds., Fluid Catalytic Cracking: Science and Technology, Studies in Surface Science and Catalysis, Vol. 76, (Amsterdam, Elsevier) 257.

Petrolite Corporation Laboratories Staff (1958). Impurities in Petroleum, Occurrence, Analysis, Significance to Refinery and Petrochemical Operations. (Houston, Petrolite Corporation).

Petrovsky, J. (1994). Question & Answer Session on Waste Issues", 1994 NPRA Refinery & Petrochemical Plant Environmental Conference (Washington, D. C., National Petroleum Refiners Association).

Pine, L. (1990). Vanadium-Catalyzed Destruction of USY Zeolites, J. Catal., 125, 514.

Pourbaix, M. (1974) Atlas of Electrochemical Equilibria in Aqueous Solutions, (Houston, TX: National Association of Corrosion Engineers).

Rosenqvist, T. (1983). Principles of Extractive Metallurgy, (New York, NY, McGraw-Hill Book Company).

Texas Corporate Recycling Council (1992), Texas Land Commission news release, re: 1992 Vision Award, Texas American Statesman, February 26.

Theodorou, D and Wei, J. (1983). "Diffusion and Reaction in Blocked and High Occupancy Zeolite Catalysts", Journal of Catalysis, 83, 205.

Trimm, D. L. (1980). <u>Design of Industrial Catalysts</u>. (New York: Elsevier), 65-70.

Yen, T. F., ed., (1975). <u>The Role of Trace Metals in Petroleum</u>, (Ann Arbor, MI: Ann Arbor Science Publishers, Inc.), 214 p.

Young, G. W. (1993). Realistic Assessment of FCC Catalyst Performance in the Laboratory. J. S. Magee and M. M. Mitchell, Jr., Eds., <u>Fluid Catalytic Cracking: Science and Technology</u>, Studies in Surface Science and Catalysis, Vol. 76, (Amsterdam, Elsevier) 257.

INDEX